SpringerBriefs in Applied Sciences and Technology

Continuum Mechanics

T0238964

For further volumes:
http://www.springer.com/series/8884

Victor A. Eremeyev · Leonid P. Lebedev
Holm Altenbach

Foundations of Micropolar Mechanics

Springer

Victor A. Eremeyev
Lehrstuhl für Technische Mechanik
Zentrum für Ingenieurwissenschaften
Martin-Luther-Universität
Halle-Wittenberg
Kurt-Mothes-Str. 1
06120 Halle (Saale)
Germany

South Federal University and South
Scientific Center of RASci
Rostov on Don
Russia

Leonid P. Lebedev
Universidad Nacional de Colombia
Cr. 45, # 26–85
Bogotá
Colombia

Holm Altenbach
Lehrstuhl für Technische Mechanik
Institut für Mechanik
Fakultät für Maschinenbau
Otto-von-Guericke-Universität Magdeburg
Universitätsplatz 2
39106 Magdeburg
Germany

ISSN 2191-530X ISSN 2191-5318 (electronic)
ISBN 978-3-642-28352-9 ISBN 978-3-642-28353-6 (eBook)
DOI 10.1007/978-3-642-28353-6
Springer Heidelberg New York Dordrecht London

Library of Congress Control Number: 2012936967

Printed on acid-free paper

Springer is part of Springer Science+Business Media (www.springer.com)

Preface

Mechanics of Micropolar Continua (also known as Cosserat Continua) is in the focus of scientists since the end of the nineteenth century. A first summary of the theory with independent force and moment (couple) actions was given in 1909 by the Cosserat brothers in their centurial book "Théorie des corps déformables" [1]. Since that time there were published tens of books and thousands of papers in this field, see e.g., the well-known books by Eringen [2, 3], Nowacki [4], and the recent collections edited by Maugin and Metrikine [5], Altenbach, Maugin and Erofeev [6], Markert [7], and Altenbach and Eremeyev [8] where the state of the art in the field of the Cosserat theory is presented. It is worth mentioning the special issue of International Journal of Engineering Science in memory of A. C. Eringen, which is published as Volume 49(2011), No. 12, edited by G. A. Maugin and J. D. Lee. All the contributions on the Cosserat continuum are focussed on the fact that the continuum translations and rotations can be defined independently. In other words, force and moment actions in the continuum can be introduced independently as in dynamics of rigid body or structural mechanics. In a micropolar medium, each material particle has six degrees of freedom, they are three translational and three rotational degrees of freedom. Besides ordinary stresses in the theory of micropolar continuum there are introduced couple stresses. These characteristic features of the Cosserat continuum model give a possibility to describe more complex media, for example, micro-inhomogeneous materials, foams, cellular solids, lattices, masonries, particle assemblies, magnetic rheological fluids, liquid crystals, etc. The aim of this book is as follows:

- the presentation of the basics of the micropolar continuum mechanics including a short but comprehensive introduction of stress and strain measures, derivation of motion equations, and discussion of the differences between Cosserat and classical (Cauchy) continua, and
- the discussion of more specific problems related to the constitutive modeling, i.e., constitutive inequalities, symmetry groups, acceleration waves, etc., which are original contributions in the field.

Chapter 1 (Introduction) is a brief review of some important publications in the research area under consideration. Such a review cannot be complete—there are too many publications. But it is a starting point for further studies. In Chap. 2 we recall kinematics of the micropolar continuum and introduce the directors and the microrotation tensor. In Chap. 3 we consider the theory of stresses and couple stresses. Here, we present the motion equations of micropolar continuum and Lagrangian and Eulerian statements of the boundary value problems. Chapter 4 is devoted to the theory of constitutive equations. We formulate the principles of determinism, local action, and material frame-indifference. Here, we consider the constitutive equations of the hyperelastic micropolar continuum. We introduce the natural relative Lagrangian strain measures using three different approaches. Then we present the definition of the material symmetry group and discuss the corresponding constitutive equations. We introduce constraints which lead for example to the model of the Cosserat continuum with the constrained rotations. We establish the condition of strong ellipticity of equilibrium equations and formulate other constitutive inequalities. The detailed analysis of acceleration waves in thermoelastic micropolar continuum is the subject of Chap. 5. We prove that the conditions of the existence of acceleration waves coincide with the condition of strong ellipticity of equilibrium equations. A few examples are presented.

The book also contains some necessary information about tensor analysis. Using Cosserat's approach we briefly present some elements of rigid body dynamics and the mechanics of rods and shells to demonstrate that Cosserat's description of continua fills up the gap between the Continuum Mechanics and Mechanics of rigid bodies and thin structures.

In what follows, we use the notations: the numeration of formulae, figures, and tables is given as (i.j), where i is the number of the chapter and j is the number inside the chapter; the vectors are denoted by semibold roman font like $\mathbf{A}, \mathbf{a}$, the second-order tensors are denoted by semibold san serif font like A, a, the higher order tensors are denoted by blackboard bold font like $\mathbb{A}$, Greece indices take values 1 and 2, whereas Latin indices are arbitrary. We also employ the Einstein summation convention.

This book is written for specialists in continuum mechanics as well as for Masters and PhD students investigating Cosserat and other generalized continua.

The authors wish to express special thanks to Prof. Andreas Öchsner (acting as the Springer Briefs series editor) and Dr. Christoph Baumann as a responsible person from Springer Publishers Group.

The first author (V.A.E.) acknowledges the support by the German Research Foundation (grant AL341/33-1) and by the Russian Foundation for Basic Research (grants 09-01-00459, 12-01-00038).

Halle (Saale), Bogotá, Magdeburg,
January, 2012

Victor A. Eremeyev
Leonid P. Lebedev
Holm Altenbach

References

1. E. Cosserat, F. Cosserat, *Théorie des corps déformables* (Herman et Fils, Paris, 1909)
2. A.C. Eringen, *Microcontinuum Field Theory I. Foundations and Solids* (Springer, New York, 1999)
3. A.C. Eringen, *Microcontinuum Field Theory II. Fluent Media* (Springer, New York, 2001)
4. W. Nowacki, *Theory of Asymmetric Elasticity* (Pergamon-Press, Oxford, 1986)
5. G.A. Maugin, A.V. Metrikine (eds.), *Mechanics of Generalized Continua: One Hundred Years After the Cosserats* (Springer, New York, 2010)
6. H. Altenbach, G.A. Maugin, V. Erofeev (eds.), *Mechanics of Generalized Continua, Advanced Structured Materials*, vol. 7 (Springer, Berlin, 2011)
7. B. Markert (ed.), *Advances in Extended and Multifield Theories for Continua, Lecture Notes in Applied and Computational Mechanics*, vol. 59 (Springer, Berlin, 2011)
8. H. Altenbach, V.A. Eremeyev (eds.), *Generalized Continua: From the Theory to Engineering Applications*, CISM Courses and Lectures (Springer, Wien, 2012)

Contents

Chapter 1
Introduction

More than one hundred years ago the Cosserat brothers, Eugène and François, published the monograph "Théorie des corps déformables" [1], where they presented a new version of Continuum Mechanics that includes the Mechanics of Rods and Shells. It was based on the idea to consider the rotational degrees of freedom of material particles as independent variables and so to add into the material description the couple stresses. This material model was later named the Cosserat (*or* micropolar) continuum. The basic ideas of this approach were first presented in [2].

At the end of 19th century the main ideas leading to the micropolar continuum and other generalized media models were discussed by Kelvin, Helmholtz, Duhem, Voigt, Le Roux, the Cosserats and others. In a micropolar continuum, each material particle has six degrees of freedom. From the physical point of view, every material point (*or* a particle) of a micropolar continuum is phenomenologically equivalent to an infinitesimal rigid body. Hence, the rotations of close particles are taken into account independently. Note that by this approach the gap between the General Mechanics and the Strength of Materials was closed by the introduction of independent translations and rotations as well as by the analogy between the forces–couples of classic mechanics and force and moment stresses of continuum mechanics. see, e.g., [3]. A first similar result was noted by Euler. Discussing one of Lagrange's papers, he established that the foundations of mechanics are based on two principles: the momentum principle and the principle of moment of momentum. The both principles result in the Euler laws of motion [3]. In [4] there is a comment: independence of the principle of moment of momentum, which is a generalization of the static equilibrium of the moments, was established by Jacob Bernoulli (1686) one year before Newton's laws (1687). It must be noted that Newton's laws give a satisfying description of the motion of material points. If the continuum is presented by material particles of arbitrary shapes it is not sufficient.

Hence, the mechanics of generalized continua such as Cosserat continua differs from the classical (*or* Cauchy-type) continuum mechanics, it is a branch of mechanics with the origins in the 17th century. However the first serious theoretical discussions were initiated in the middle of the 19th century. The micropolar theory is of the

V. A. Eremeyev et al., *Foundations of Micropolar Mechanics*,
SpringerBriefs in Continuum Mechanics,
DOI: 10.1007/978-3-642-28353-6_1, © The Author(s) 2013

current and emerging interest of mechanicians, physicians, materials scientists as well as engineers as its limits and possibilities are not fully established that is a serious constraint for applications.

At the same time another tendency in Mechanics can be observed. Since Lagrange's "Mécanique analytique" (1788) summarizing the state of the art in Mechanics at the end of the 18th century, Mechanics splits into two branches: the mathematical and engineering ones [5]. As a result, the mathematical branch is developed more axiomatically while the engineering branch is focussed on technical applications. The examples of axiomatic approach are given by Herglotz' Lectures [6] or by the Cosserats' monograph [1]. The necessity to establish axiomatic foundations of Mechanics was pointed out by Hilbert in 1900 at the 2nd International Congress of Mathematicians in Paris. The 6th Hilbert problem states the demand to establish axiomatic structures of the Theory of Probability and Mechanics (now extended to Physics). During the first half of the 20th century only few scientist worked on this problem. One of them—Hamel—published his results since 1908 in [7], see also [8]. Since the 1950s the interest to the axiomatic approaches and Cosserats' ideas is growing again, it resulted in many publications. One of the initiators of this direction was Noll, see [9].

As it was mentioned, in addition to the ordinary stresses, in the micropolar theory the couple stresses are introduced, see [10, 11]. Deformation of the micropolar continuum can be described by the position vector $\mathbf{r}$ and three orthonormal vectors $\mathbf{d}_i$, $i = 1, 2, 3$, so-called *directors* that present the translations and the orientation changes of material particles. Introducing the directors $\mathbf{D}_k$ in a reference configuration, we can use the orthogonal tensor $\mathbf{Q} = \mathbf{d}_k \otimes \mathbf{D}_k$ for the description of relative rotations. Small deformations of the micropolar medium are defined by two independent fields: the translation vector field $\mathbf{u}$ and the microrotation vector field $\boldsymbol{\vartheta}$. The linear Cosserat theory is developed in the original papers by Günther [12], Aero and Kuvshinskii [13, 14], Toupin [15], Mindlin and Tiersten [16], Koiter [17], Palmov [18], Eringen [19, 20], Schaefer [21], Ieşan [22] among others. The problem of finite deformation of the micropolar continuum is considered by Grioli [23, 24], Toupin [25], Green and Rivlin [26], Kafadar and Eringen [27, 28], Stojanović [29–31], Besdo [32], and Reissner [33–35]. In 1960s there were published various results on the Cosserat continuum. Stojanović [29, 30] collects more than 500 references in the field. The first set of papers on generalized continua and Cosserat media is given in [36–38].

We should also mention some recent books on the Cosserat continuum [39–49], where many references to other papers can be found. It is worth of mentioning such original papers as [50–64], where the micropolar elasticity is considered, and the collected papers in [65–68].

Within the Cosserat continuum theory, many problems were successfully solved. They demonstrate qualitative and quantitative difference between the solutions by micropolar and classic elasticity models. One of the principal difficulties of any micropolar theory is to establish its constitutive equations. This question was not discussed in the Cosserats' original monograph [1], and it was a reason why the ideas of micropolar continuum were not recognized by many researchers. But even if for a material the constitutional equations are formulated we are faced with a new hard problem: the identification of the material parameters. For a homogeneous material in classical elasticity we should know two Lamé moduli whereas in the linear micropolar elasticity of isotropic solids the material is determined by six material parameters. The direct experimental identification of elastic moduli is discussed in a few papers, see, for example, the data in [42, 44]. Another approach to the determination of the micropolar moduli is based on various homogenization procedures. The third approach to the constitutive modeling of the Cosserat medium relates to the numerics. For example, the Cosserats' description is used as the base to construct special enhanced finite elements (Cosserat ponts) or as special regularization tools.

The Cosserats continuum model is widely applied to the description of plastic behavior of solids as well. One of the first such papers on plasticity is the one by Lippmann [69], where the yield criterium is generalized taking into account the couple stresses. Cosserat plasticity appears in many recent works see the papers by de Borst [70, 71], de Borst et al. [72, 73], Ehlers [74], Forest [75, 76], Forest et al. [77–82], Grammenoudis and Tsakmakis [83–85], Neff [86], Neff and Chełmiński [87, 88], Steinmann [89, 90], Steinmann et al. [62, 91, 92], and [93–97]. The idea of applying of Cosserats' approach to the description of plasticity effects seems to be quite natural as the approach presents a possibility to take into account the independence of grain rotations in polycrystalline materials during plastic deformation.

The Cosserat model is used to describe solid materials with a complex microstructure like soils, polycrystalline and composite materials, granular and powder-like materials, masonries, see the papers by Besdo [98], Bigoni and Drugan [99], Diebels [100, 101], Dos Reis and Ganghoffer [102], Ehlers et al. [74, 95, 103], Forest et al. [79, 104, 105], Suiker et al. [106, 107], and [86, 108–118]. In addition, the model can be applied to nanostructures [119], cellular or porous media and foams [74, 100, 101, 120–125], and even to bones [125–127], as well as to electromagnetic and ferromagnetic media, see [114, 128, 129]. Application of the Cosserat continuum model to these media is also natural. For example, open-cell and closed-cell foams can be considered as a system of elastic beams or plates, respectively. For beams and plates the force and couple interactions are essential, so stresses and couple stresses are the primary quantities as well as in the mechanics of structures. Replacing such highly inhomogeneous material as a foam by a quasi-homogeneous medium one can expect that this medium inherits the couple stress properties and rotational freedom degrees of the initial system of beams or plates. Thus applying some homogenization techniques for a micro-inhomogeneous media one obtains an "effective" quasi-homogeneous Cosserat-type continuum. The transition from an inhomogeneous media to a generalized quasi-homogeneous one is analogous to the transition from the equations of three-dimensional (3D) elasticity to the two-dimensional (2D) equations of the shell theories with resultant stresses and couple stresses using the through-the-thickness integration technique or other 3D–to–2D reduction procedures. For an electromagnetic medium one may take into account the couples induced by the electromagnetic field that leads to the appearance of the couple stresses in the medium.

Starting with the landmark papers by Aero et al. [130] and Eringen [131], the micropolar continuum is applied to the model of fluids. This branch of the hydrodynamics is called the micropolar *or* asymmetric hydrodynamics. The micropolar hydrodynamics is applied to describe the behavior of magnetic liquids, polymer suspensions, liquid crystals, and other types of fluids with microstructure, see, for example, the books by Migoun and Prokhorenko [132], Łukaszewicz [47], and Eringen [43].

Since the paper by Ericksen and Truesdell [133], the Cosserat model was used to develop various generalized models for beams, plates, and shells. With the use of the direct approach in [133], the shell is modeled as a deformable surface, at each point of which a set of deformable directors is attached. Here the shell deformation is described by the position vector $\mathbf{r}$ and p directors $\mathbf{d}_i$, $i = 1, \ldots p$. This variant of the shell theory is also named Cosserat shell theory *or* the theory of Cosserat surfaces. We should also mention the books by Naghdi [134], Rubin [135], Antman [136], and Bîrsan [137], where the theory of Cosserat shells is presented, and the review by Altenbach et al. [138], where the different Cosserat-based approaches in shell theories are compared. In the recent literature there are the shell theories that are based on

1. the introduction of one or more deformable directors as additional kinematic variables,
2. the use of the attached three rigid (non-deformable) orthogonal directors, and
3. the reduction of three-dimensional (3D) boundary-value problems for 3D shell-like bodies to 2D equations of shells, see [138].

In particular, the micropolar shell can be described as a deformable surface with attached tree rigid directors. Such a model can be considered as a 2D micropolar continuum, many notions of 3D micropolar theory can be transferred to the 2D case.

Let us note that nowadays the Cosserat continuum has taken a significant place in Continuum Mechanics, see for example recent books by Besson et al. [139], Clayton [140], Maugin [141, 142], and Narasimhan [143], where the Cosserat model is included within the context of the mechanics of materials.

References

1. E. Cosserat, F. Cosserat, *Théorie des corps déformables* (Herman et Fils, Paris, 1909)
2. E. Cosserat, F. Cosserat, Sur la théorie de l'elasticité. Ann. Toulouse **10**, 1–116 (1896)
3. C. Truesdell, Die Entwicklung des Drallsatzes. ZAMM **44**(4/5), 149–158 (1964)
4. G. Naue, Kontinuumsbegriff und Erhaltungssätze in der Mechanik seit Leonard Euler. Technische Mechanik **5**(4), 62–66 (1984)
5. H. Altenbach, Zu einigen Aspekten der klassischen Kontinuumsmechanik und ihrer Erweiterungen. Technische Mechanik **11**(2), 95–105 (1990)
6. G. Herglotz, *Vorlesungen über Mechanik der Kontinua, Teubner-Archiv zur Mathematik*, vol. 3 (B.G. Teubner, Leipzig, 1985)
7. G. Hamel, *Elementare Mechanik: ein Lehrbuch enthaltend: eine Begründung der allgemeinen Mechanik, die Mechanik der Systeme starrer Körper, die synthetischen und die Elemente der*

analytischen Methoden, sowie eine Einführung in die Prinzipien der Mechanik deformierbarer Systeme (Teubner, Leipzig, 1912)
8. G. Hamel, *Theoretische Mechanik: eine einheitliche Einführung in die gesamte Mechanik* (Springer, Berlin, 1949)
9. B.D. Coleman, M. Feinberg, J. Serrin (eds.), *Analysis and Thermodynamics—A Collection of Papers Dedicated to W. Noll* (Springer, Berlin, 1987)
10. C. Truesdell, W. Noll, The nonlinear field theories of mechanics, in *Handbuch der Physik*, vol. III/3, ed. by S. Flügge (Springer, Berlin, 1965), pp. 1–602
11. C. Truesdell, R. Toupin, The classical field theories, in *Handbuch der Physik*, vol. III/1, ed. by S. Flügge (Springer, Berlin, 1960), pp. 226–793
12. W. Günther, Zur Statik und Kinematik des Cosseratschen Kontinuums. Abhandlungen der Braunschweigschen Wissenschaftlichen Gesellschaft, Göttingen **10**, 196–213 (1958)
13. E.L. Aero, E.V. Kuvshinskii, Fundamental equations of the theory of elastic media with rotationally interacting particles. Sov. Phys. Solid State **2**(7), 1272–1281 (1961)
14. E.L. Aero, E.V. Kuvshinskii, Continuum theory of asymmetric elasticity. Equilibrium of an isotropic body (in Russian). Fizika Tverdogo Tela **6**, 2689–2699 (1964)
15. R.A. Toupin, Elastic materials with couple-stresses. Arch. Ration. Mech. Anal. **11**, 385–414 (1962)
16. R.D. Mindlin, H.F. Tiersten, Effects of couple-stresses in linear elasticity. Arch. Ration. Mech. Anal. **11**, 415–448 (1962)
17. W.T. Koiter, Couple-stresses in the theory of elasticity. Pt I-II. Proc. Koninkl. Neterland. Akad. Wetensh. B **67**, 17–44 (1964)
18. V.A. Pal'mov, Fundamental equations of the theory of asymmetric elasticity. J. Appl. Mech. Math. **28**(3), 496–505 (1964)
19. A.C. Eringen, Linear theory of micropolar elasticity. J. Math. Mech. **15**(6), 909–923 (1966)
20. A.C. Eringen, Linear theory of micropolar viscoelasticity. Int. J. Eng. Sci. **5**(2), 191–204 (1967)
21. H. Schaefer, Das Cosserat-Kontinuum. ZAMM **47**(8), 485–498 (1967)
22. D. Iesan, On the linear theory of micropolar elasticity. Intl. J. Eng. Sci. **7**(12), 1213–1220 (1969)
23. G. Grioli, Elasticita asimmetrica. Annali di Matematica Pura ed Applicata **50**(1), 389–417 (1960)
24. G. Grioli, Contributo per una formulazione di tipo integrale della meccanica dei continui di Cosserat. Annali di Matematica Pura ed Applicata **111**(1), 389–417 (1976)
25. R.A. Toupin, Theories of elasticity with couple-stress. Arch. Ration. Mech. Anal. **17**, 85–112 (1964)
26. A.E. Green, R.S. Rivlin, Multipolar continuum mechanics. Arch. Ration. Mech. Anal. **17**(2), 113–147 (1964)
27. C.B. Kafadar, A.C. Eringen, Micropolar media - I. The classical theory. Int. J. Eng. Sci. **9**(3), 271–305 (1971)
28. A.C. Eringen, C.B. Kafadar, Polar field theories, in *Continuum Physics*, vol. IV, ed. by A.C. Eringen (Academic Press, New York, 1976), pp. 1–75
29. R. Stojanovic, *Mechanics of Polar Continua: Theory and Applications, CISM Courses and Lectures*, vol. 2 (Springer, Wien, 1969)
30. R. Stojanovic, *Recent Developments in the Theory of Polar Continua*, CISM Courses and Lectures No. 27, vol. 27. (Springer, Wien, 1969)
31. R. Stojanovic, Nonlinear micropolar elasticity, in *Micropolar Elasticity*, vol. 151, ed. by W. Nowacki, W. Olszak (Springer, Wien, 1974), pp. 73–103
32. D. Besdo, Ein Beitrag zur nichtlinearen Theorie des Cosserat-Kontinuums. Acta Mechanica **20**(1–2), 105–131 (1974)
33. E. Reissner, On kinematics and statics of finite-strain force and moment stress elasticity. Stud. Appl. Math. **52**, 93–101 (1973)
34. E. Reissner, Note on the equations of finite-strain force and moment stress elasticity. Stud. Appl. Math. **54**, 1–8 (1975)

35. E. Reissner, A further note on the equations of finite-strain force and moment stress elasticity. ZAMP **38**, 665–673 (1987)
36. E. Kröner, (ed.), *Mechanics of Generalized Continua. Proceedings of the IUTAM-Symposium on the Generalized Cosserat Continuum and the Continuum Theory of Dislocations with Applications, Freudenstadt and Stuttgart (Germany), 1967*, vol. 16. (Springer, Berlin, 1968)
37. W. Nowacki (ed.), *Theory of Micropolar Elasticity, CISM Courses and Lectures*, vol. 25 (Springer, Wien, 1970)
38. W. Nowacki, W. Olszak (eds.), *Micropolar Elasticity, CISM Courses and Lectures*, vol. 151 (Springer, Wien, 1974)
39. G. Capriz, *Continua with Microstructure* (Springer, New York, 1989)
40. M. Ciarletta, D. Iesan, *Non-classical Elastic Solids* (Longman, UK, 1993)
41. J. Dyszlewicz, *Micropolar Theory of Elasticity* (Springer, Berlin, 2004)
42. A.C. Eringen, *Microcontinuum Field Theory I. Foundations and Solids* (Springer, New York, 1999)
43. A.C. Eringen, *Microcontinuum Field Theory II. Fluent Media* (Springer, New York, 2001)
44. V.I. Erofeev, *Wave Processes in Solids with Microstructure* (World Scientific, Singapore, 2003)
45. D. Iesan, *Saint-Venant's Problem* (Springer, Berlin, 1987)
46. D. Iesan, *Classical and Generalized Models of Elastic Rods* (CRC Press, Boca Raton, 2009)
47. G. Lukaszewicz, *Micropolar Fluids: Theory and Applications* (Birkhäuser, Boston, 1999)
48. W. Nowacki, *Theory of Asymmetric Elasticity* (Pergamon-Press, Oxford, 1986)
49. L.M. Zubov, *Nonlinear Theory of Dislocations and Disclinations in Elastic Bodies* (Springer, Berlin, 1997)
50. P.H. Dłuzevski, Finite deformations of polar elastic media. Int. J. Solids Struct. **30**(16), 2277–2285 (1993)
51. V.A. Eremeyev, L.M. Zubov, On the stability of elastic bodies with couple stresses. Mech. Solids **29**(3), 172–181 (1994)
52. E. Grekova, P. Zhilin, Basic equations of Kelvin's medium and analogy with ferromagnets. J. Elast. **64**, 29–70 (2001)
53. G.A. Maugin, On the structure of the theory of polar elasticity. Philos. Trans. Royal Soc. Lond. A **356**(1741), 1367–1395 (1998)
54. P. Neff, The Cosserat couple modulus for continuous solids is zero viz the linearized Cauchy-stress tensor is symmetric. ZAMM **86**(11), 892–912 (2006)
55. J. Jeong, P. Neff, Existence, uniqueness and stability in linear Cosserat elasticity for weakest curvature conditions. Math. Mech. Solids **15**(1), 78–95 (2010)
56. P. Neff, J. Jeong, A new paradigm: the linear isotropic Cosserat model with conformally invariant curvature energy. ZAMM **89**(2), 107–122 (2009)
57. I. Nistor, Variational principles for Cosserat bodies. Int. J. Non-Linear Mech. **37**(3), 565–569 (2002)
58. E. Nikitin, L.M. Zubov, Conservation laws and conjugate solutions in the elasticity of simple materials and materials with couple stress. J. Elast. **51**(1), 1–22 (1998)
59. W. Pietraszkiewicz, V.A. Eremeyev, On natural strain measures of the non-linear micropolar continuum. Int. J. Solids Struct. **46**(3–4), 774–787 (2009)
60. S. Ramezani, R. Naghdabadi, Energy pairs in the micropolar continuum. Int. J. Solids Struct. **44**(14–15), 4810–4818 (2007)
61. S. Ramezani, R. Naghdabadi, S. Sohrabpour, Constitutive equations for micropolar hyper-elastic materials. Int. J. Solids Struct. **46**(14–15), 2765–2773 (2009)
62. P. Steinmann, E. Stein, A unifying treatise of variational principles for two types of micropolar continua. Acta Mechanica **121**, 215–232 (1997)
63. H. Xiao, On symmetries and anisotropies of classical and micropolar linear elasticities: a new method based upon a complex vector basis and some systematic results. J. Elast. **49**(2), 129–162 (1998)
64. L.M. Zubov, Continuously distributed dislocations and disclinations in nonlinearly elastic micropolar media. Doklady Phys. **49**(5), 308–310 (2004)

65. G.A. Maugin, A.V. Metrikine (eds.), *Mechanics of Generalized Continua: One Hundred Years After the Cosserats* (Springer, New York, 2010)
66. H. Altenbach, G.A. Maugin, V. Erofeev (eds.), *Mechanics of Generalized Continua, Advanced Structured Materials*, vol. 7 (Springer, Berlin, 2011)
67. B. Markert (ed.), *Advances in Extended and Multifield Theories for Continua, Lecture Notes in Applied and Computational Mechanics*, vol. 59 (Springer, Berlin, 2011)
68. H. Altenbach, V.A. Eremeyev (eds.), *Generalized Continua: From the Theory to Engineering Applications*, CISM Courses and Lectures, (Springer, Wien, 2012)
69. H. Lippmann, Eine Cosserat-Theorie des plastischen Fließens. Acta Mechanica **8**(3–4), 93–113 (1969)
70. R. de Borst, Simulation of strain localization: a reappraisal of the Cosserat continuum. Eng. Comput. **8**(4), 317–332 (1991)
71. R. de Borst, A generalisation of J_2-flow theory for polar continua. Comput. Methods Appl. Mech. Eng. **103**(3), 347–362 (1993)
72. R. de Borst, H.B. Mühlhaus, Gradient-dependent plasticity: formulation and algorithmic aspects. Int. J. Numer. Methods Eng. **35**(3), 521–539 (1992)
73. R. de Borst, L.J. Sluys, H.B. Mühlhaus, J. Pamin, Fundamental issues in finite element analyses of localization of deformation. Eng. Comput. **10**(2), 99–121 (1993)
74. W. Ehlers, Theoretical and numerical modelling of granular liquid-saturated elasto-plastic porous media. ZAMM **77**(Suppl. 2), S401–S404 (1997)
75. S. Forest, Modeling slip, kink and shear banding in classical and generalized single crystal plasticity. Acta Materialia **46**(9), 3265–3281 (1998)
76. S. Forest, Some links between Cosserat, strain gradient crystal plasticity and the statistical theory of dislocations. Philos. Mag. **88**(30–32), 3549–3563 (2008)
77. S. Forest, R. Sedlávcek, Plastic slip distribution in two-phase laminate microstructures: dislocation-based versus generalized continuum approaches. Philos. Mag. **83**(2), 245–276 (2003)
78. S. Forest, R. Sievert, Elastoviscoplastic constitutive frameworks for generalized continua. Acta Mechanica **160**(1–2), 71–111 (2003)
79. S. Forest, R. Sievert, Nonlinear microstrain theories. Int. J. Solids Struct. **43**(24), 7224–7245 (2006)
80. S. Forest, F. Barbe, G. Cailletaud, Cosserat modelling of size effects in the mechanical behaviour of polycrystals and multi-phase materials. Int. J. Solids Struct. **37**(46–47), 7105–7126 (2000)
81. R. Sedlávcek, S. Forest, Non-local plasticity at microscale: a dislocation-based and a Cosserat model. Phys. Status Solidi B. Basic Res. **221**(2), 583–596 (2000)
82. R. Sievert, S. Forest, R. Trostel, Finite deformation Cosserat-type modelling of dissipative solids and its application to crystal plasticity. J. de Physique IV **8**(P8), 357–364 (1998).
83. P. Grammenoudis, C. Tsakmakis, Finite element implementation of large deformation micropolar plasticity exhibiting isotropic and kinematic hardening effects. Int. J. Numer. Methods Eng. **62**(12), 1691–1720 (2005)
84. P. Grammenoudis, C. Tsakmakis, Predictions of microtorsional experiments by micropolar plasticity. Proc. Royal Soc. Lond. A **461**(2053), 189–205 (2005)
85. P. Grammenoudis, C. Tsakmakis, Isotropic hardening in micropolar plasticity. Arch. Appl. Mech. **79**(4), 323–334 (2009)
86. P. Neff, A finite-strain elastic-plastic Cosserat theory for polycrystals with grain rotations. Int. J. Eng. Sci. **44**(8–9), 574–594 (2006)
87. K. Chełminski, P. Neff, A note on approximation of Prandtl-Reuss plasticity through Cosserat plasticity. Q. Appl. Math. **66**(2), 351–357 (2008)
88. Neff, P., Chełminski, K.: Infinitesimal elastic-plastic Cosserat micropolar theory. Modelling and global existence in the rate independent case. Proc. Royal Soc. Edinb. A. Math. **135**(5), 1017–1039 (2005).
89. P. Steinmann, Theorie endlicher mikropolarer Elasto-Plastizität. ZAMM **74**(4), T245–T247 (1994)

90. P. Steinmann, A micropolar theory of finite deformation and finite rotation multiplicative elastoplasticity. Int. J. Solids Struct. **31**(8), 1063–1084 (1994)
91. A. Dietsche, P. Steinmann, K. Willam, Micropolar elastoplasticity and its role in localization. Int. J. Plast. **9**(7), 813–831 (1993)
92. G. Etse, M. Nieto, P. Steinmann, A micropolar microplane theory. Int. J. Eng. Sci. **41**(13–14), 1631–1648 (2003)
93. D. Peric, J. Yu, D.R.J. Owen, On error estimates and adaptivity in elastoplastic solids: applications to the numerical simulation of strain localization in classical and Cosserat continua. Int. J. Numer. Methods Eng. **37**(8), 1351–1379 (1994)
94. A. Dietsche, K. Willam, Boundary effects in elasto-plastic Cosserat continua. Int. J. Solids Struct. **34**(7), 877–893 (1997)
95. W. Ehlers, W. Volk, On shear band localization phenomena of liquid-saturated granular elastoplastic porous solid materials accounting for fluid viscosity and micropolar solid rotations. Mech. Cohesive-Frict. Mater. **2**(4), 301–320 (1997)
96. P. Neff, K. Chełminski, W. Müller, C. Wieners, A numerical solution method for an infinitesimal elasto-plastic Cosserat model. Math. Models Methods Appl. Sci. **17**(8), 1211–1239 (2007)
97. M. Ristinmaa, M. Vecchi, Use of couple-stress theory in elasto-plasticity. Comput. Methods Appl. Mech. Eng. **136**(3–4), 205–224 (1996)
98. D. Besdo, Towards a Cosserat-theory describing motion of an originally rectangular structure of blocks. Arch. Appl. Mech. **80**(1), 25–45 (2010)
99. D. Bigoni, W.J. Drugan, Analytical derivation of Cosserat moduli via homogenization of heterogeneous elastic materials. Trans. ASME J. Appl. Mech. **74**(4), 741–753 (2007)
100. S. Diebels, A micropolar theory of porous media: constitutive modelling. Trans. Porous Media **34**(1–3), 193–208 (1999)
101. S. Diebels, A macroscopic description of the quasi-static behavior of granular materials based on the theory of porous media. Granular Matter **2**(3), 143–152 (2000)
102. F. Dos Reis, J.F. Ganghoffer, Construction of micropolar continua from the homogenization of repetitive planar lattices, in *Mechanics of Generalized Continua, Advanced Structured Materials*, vol. 7, ed. by H. Altenbach, G.A. Maugin, V. Erofeev (Springer, Berlin, 2011), pp. 193–217
103. W. Ehlers, E. Ramm, S. Diebels, G.D.A. d'Addetta, From particle ensembles to Cosserat continua: Homogenization of contact forces towards stresses and couple stresses. Int. J. Solids Struct. **40**(24), 6681–6702 (2003)
104. Forest, S.: Mechanics of generalized continua: construction by homogenizaton. J. de Physique IV France **8**(PR4), Pr4-39–Pr4-48 (1998)
105. S. Forest, K. Sab, Cosserat overal modelling of heterogeneous materials. Mech. Res. Commun. **25**(4), 449–454 (1998)
106. A.S.J. Suiker, A.V. Metrikine, R. de Borst, Comparison of wave propagation characteristics of the Cosserat continuum model and corresponding discrete lattice models. Int. J. Solids Struct. **38**(9), 1563–1583 (2001)
107. A.S.J. Suiker, R. de Borst, Enhanced continua and discrete lattices for modelling granular assemblies. Philos. Trans. Royal Soc. A **363**(1836), 2543–2580 (2005)
108. R. Larsson, S. Diebels, A second-order homogenization procedure for multi-scale analysis based on micropolar kinematics. Int. J. Numer. Methods Eng. **69**(12), 2485–2512 (2007)
109. R. Larsson, Y. Zhang, Homogenization of microsystem interconnects based on micropolar theory and discontinuous kinematics. J. Mech. Phys. Solids **55**(4), 819–841 (2007)
110. E. Pasternak, H.B. Mühlhaus, Generalised homogenisation procedures for granular materials. J. Eng. Math. **52**(1), 199–229 (2005)
111. O. van der Sluis, P.H.J. Vosbeek, P.J.G. Schreurs, H.E.H. Meijer, Homogenization of heterogeneous polymers. Int. J. Solids Struct. **36**(21), 3193–3214 (1999)
112. P. Trovalusci, R. Masiani, Non-linear micropolar and classical continua for anisotropic discontinuous materials. Int. J. Solids Struct. **40**(5), 1281–1297 (2003)

113. G. Capriz, P. Giovine, P.M. Mariano (eds.), *Mathematical Models of Granular Matter* (Springer, Berlin, 2008)
114. H. Kotera, M. Sawada, S. Shima, Cosserat continuum theory to simulate microscopic rotation of magnetic powder in applied magnetic field. Int. J. Mech. Sci. **42**(1), 129–145 (2000)
115. T. Matsushima, H. Saomoto, Y. Tsubokawa, Y. Yamada, Grain rotation versus continuum rotation during shear deformation of granular assembly. Soils Found. **43**(4), 95–106 (2003)
116. K. Mori, M. Shiomi, K. Osakada, Inclusion of microscopic rotation of powder particles during compaction in finite element method using Cosserat continuum theory. Int. J. Numer. Methods Eng. **42**(5), 847–856 (1998)
117. J. Tejchman, *Shear Localization in Granular Bodies with Micro-Polar Hypoplasticity* (Springer, Berlin, 2008)
118. H.W. Zhang, H. Wang, P. Wriggers, B.A. Schrefler, A finite element model for contact analysis of multiple Cosserat bodies. Comput. Mech. **36**(6), 444–458 (2005)
119. E.A. Ivanova, A.M. Krivtsov, N.F. Morozov, A.D. Firsova, Description of crystal packing of particles with torque interaction. Mech. Solids **38**(4), 76–88 (2003)
120. S. Diebels, H. Steeb, The size effect in foams and its theoretical and numerical investigation. Proc. Royal Soc. Lond. A **458**(3), 2869–2883 (2002)
121. S. Diebels, H. Steeb, Stress and couple stress in foams. Comput. Mater. Sci. **28**(3–4), 714–722 (2003)
122. T. Dillard, S. Forest, P. Ienny, Micromorphic continuum modelling of the deformation and fracture behaviour of nickel foams. Europ. J. Mech. A-Solids **25**(3), 526–549 (2006)
123. R.S. Lakes, Experimental microelasticity of two porous solids. Int. J. Solids Struct. **22**(1), 55–63 (1986)
124. R.S. Lakes, Experimental micro mechanics methods for conventional and negative Poisson's ratio cellular solids as Cosserat continua. Trans. ASME J. Eng. Mater. Technol. **113**(1), 148–155 (1991)
125. R.S. Lakes, Experimental methods for study of Cosserat elastic solids and other generalized continua. in *Continuum Models for Materials with Micro-Structure*, ed. by H. Mühlhaus (Wiley, New York, 1995) pp. 1–22
126. J.F.C. Yang, R.S. Lakes, Experimental study of micropolar and couple stress elasticity in compact bone in bending. J. Biomech. **15**(2), 91–98 (1982)
127. H.C. Park, R.S. Lakes, Cosserat micromechanics of human bone: strain redistribution by a hydration-sensitive constituent. J. Biomech. **19**(5), 385–397 (1986)
128. A.C. Eringen, G.A. Maugin, *Electrodynamics of Continua* (Springer, New York, 1990)
129. G.A. Maugin, *Continuum Mechanics of Electromagnetic Solids* (Elsevier, Oxford, 1988)
130. E.L. Aero, A.N. Bulygin, E.V. Kuvshinskii, Asymmetric hydromechanics. J. Appl. Math. Mech. **29**(2), 333–346 (1965)
131. A.C. Eringen, Theory of micropolar fluids. J. Math. Mech. **16**(1), 1–18 (1966)
132. N.P. Migoun, P.P. Prokhorenko, *Hydrodynamics and Heattransfer in Gradient Flows of Microstructured Fluids (in Russian)* (Nauka i Technika, Minsk, 1984)
133. J.L. Ericksen, C. Truesdell, Exact tbeory of stress and strain in rods and shells. Arch. Ration. Mech. Anal. **1**(1), 295–323 (1958)
134. P. Naghdi, The theory of plates and shells, in *Handbuch der Physik*, vol. VIa/2, ed. by S. Flügge (Springer, Heidelberg, 1972), pp. 425–640
135. M.B. Rubin, *Cosserat Theories: Shells Rods and Points* (Kluwer, Dordrecht, 2000)
136. S.S. Antman, *Nonlinear Problems of Elasticity*, 2nd edn. (Springer Science Media, New York, 2005)
137. M. Bîrsan, *Linear Cosserat Elastic Shells: Mathemathical Theory and Applications* (Matrix Rom, Bucuresti, 2009)
138. J. Altenbach, H. Altenbach, V.A. Eremeyev, On generalized Cosserat-type theories of plates and shells. A short review and bibliography. Arch. Appl. Mech. **80**(1), 73–92 (2010)
139. J. Besson, G. Cailletaud, J.L. Chaboche, S. Forest, M. Blétry, *Non-Linear Mechanics of Materials, Solid Mechanics and Its Applications*, vol. 167 (Springer, Dordrecht, 2010)

140. J.D. Clayton, *Nonlinear Mechanics of Crystals, Solid Mechanics and Its Applications*, vol. 177 (Springer, Dordrecht, 2011)
141. G.A. Maugin, *Nonlinear Waves on Elastic Crystals* (Oxford University Press, Oxford, 1999)
142. G.A. Maugin, *Configurational Forces: Thermomechanics, Physics, Mathematics, and Numerics* (CRC Press, Boca Raton, 2010)
143. M.N.L. Narasimhan, *Principles of Continuum Mechanics* (Wiley, New York, 1993)

Chapter 2
Kinematics of Micropolar Continuum

In this chapter we briefly recall general kinematical relations for a micropolar continuum. For a comprehensive approach, we refer the reader to [1, 2]. The symbolic (direct) tensor notation follows the one by [3, 4], see also Appendix A.

The description of motion of a particle of a micropolar continuum (medium) is based on the assumption that every particle of the micropolar body has six degrees of freedom, see [1, 2]. This is similar to the description of a rigid body in classical mechanics. Three of the degrees of freedom are translational as in classic elasticity, and other three degrees are *orientational* or *rotational*.

In the actual configuration χ at instant t, the position of a particle of micropolar continuum is given by the position vector $\mathbf{r}$. The particle orientation is defined by an orthonormal trihedron $\mathbf{d}_k$ ($k = 1, 2, 3$) whose vectors are called *directors*. The two vector fields $\mathbf{r}$ and $\mathbf{d}_k$ define the translational and rotational motions of a particle.

To describe the medium relative deformation, we use some fixed position of the body that may be taken at $t = 0$ or another fixed instant; we call this position the *reference configuration* κ. Here the state of particle is defined by the position vector $\mathbf{R}$, whereas its orientation by directors $\mathbf{D}_k$ (cf. Fig. 2.1). Let us note that as the reference configuration can be chosen not only the real state but also any one.

The motion of a micropolar continuum can be described by the following vectorial fields

$$\mathbf{r} = \mathbf{r}(\mathbf{R}, t), \quad \mathbf{d}_k = \mathbf{d}_k(\mathbf{R}, t). \tag{2.1}$$

In the process of deformation the trihedron $\mathbf{d}_k$ stays orthonormal, $\mathbf{d}_k \cdot \mathbf{d}_m = \delta_{km}$. The change of the directors can be described by an orthogonal tensor that is

$$\mathsf{H} = \mathbf{d}_k \otimes \mathbf{D}_k.$$

H is called the *microrotation tensor*. So $\mathbf{r}$ describes the position of the particle of the continuum at time t, whereas H defines its orientation. The orientation of

V. A. Eremeyev et al., *Foundations of Micropolar Mechanics*,
SpringerBriefs in Continuum Mechanics,
DOI: 10.1007/978-3-642-28353-6_2, © The Author(s) 2013

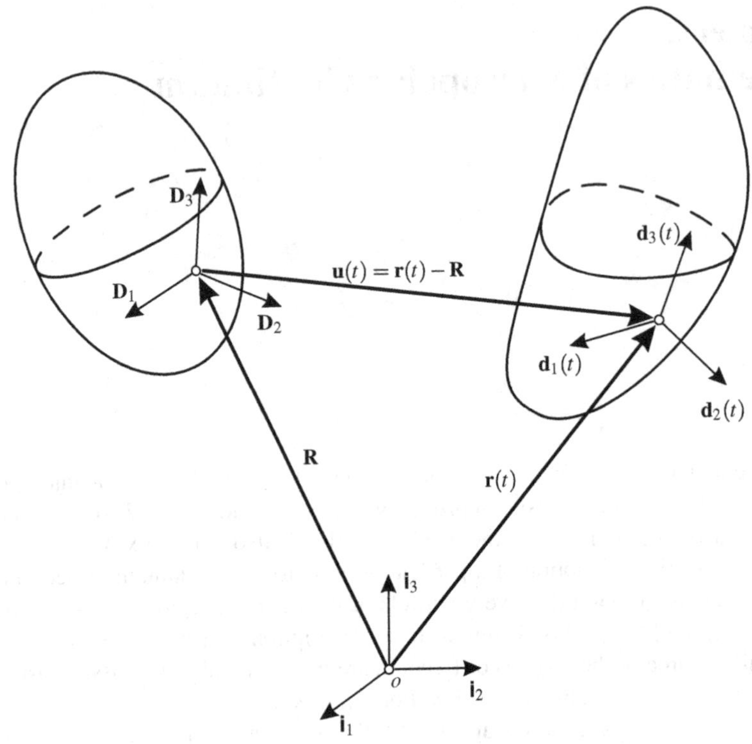

Fig. 2.1 Deformation of a micropolar body (reference and actual configurations)

$\mathbf{D}_k$ and $\mathbf{d}_k$ can be selected the same, so $\mathbf{H}$ is proper orthogonal. Hence, the micropolar continuum deformation can be described by the following relations

$$\mathbf{r} = \mathbf{r}(\mathbf{R}, t), \quad \mathbf{H} = \mathbf{H}(\mathbf{R}, t). \tag{2.2}$$

The linear velocity is given by the relation

$$\mathbf{v} = \dot{\mathbf{r}}. \tag{2.3}$$

For brevity, we use the notation $\dot{(...)}\equiv \dfrac{\mathrm{d}}{\mathrm{d}t}(...)$, where $\dfrac{\mathrm{d}}{\mathrm{d}t}$ denotes the material derivative with respect to t. As in classical mechanics, see (B.9), the angular velocity vector, called *microgyration vector*, is given by

$$\boldsymbol{\omega} = -\frac{1}{2}\left(\mathbf{H}^T \cdot \dot{\mathbf{H}}\right)_\times, \tag{2.4}$$

where the dot denotes the dot (inner) product and $(\ldots)^T$-transposed. The symbol $(...)_\times$ stands for the vector invariant of a second-order tensor (cf. (A.4)). In particular, for a dyad $\mathbf{a} \otimes \mathbf{b}$ we have $(\mathbf{a} \otimes \mathbf{b})_\times = \mathbf{a} \times \mathbf{b}$, where $\times$ is the vector (cross) product. Relation (2.4) means that $\boldsymbol{\omega}$ is the axial vector associated with the skew-symmetric tensor $\mathbf{H}^T \cdot \dot{\mathbf{H}}$.

References

1. A.C. Eringen, C.B. Kafadar, Polar field theories. in *Continuum Physics*, vol. IV, ed. by A.C. Eringen (Academic Press, New York, 1976), pp. 1–75
2. A.C. Eringen, *Microcontinuum Field Theory. I. Foundations and Solids* (Springer, New York, 1999)
3. L.P. Lebedev, M.J. Cloud, V.A. Eremeyev, *Tensor Analysis with Applications in Mechanics* (World Scientific, New Jersey, 2010)
4. A.I. Lurie, *Theory of Elasticity* (Springer, Berlin, 2005)

Chapter 3
Forces and Couples, Stress and Couple Stress Tensors in Micropolar Continua

In this chapter using the balance of momentum and balance of moment of momentum (Euler's laws of motion) we introduce the stress and couple stress tensors. Then we derive the motion equations of the micropolar continuum which contains the motion equations of simple (non-polar) continuum as a special case.

3.1 Forces and Couples

Forces and *couples* are the primary quantities of continuum mechanics. It is possible to introduce them in the axiomatic way as it was done by Truesdell and Noll for simple materials [1–3].

Let us consider an arbitrary part $\mathscr{P}$ of a material body $\mathscr{B}$ at instant t, cf. Fig. 3.1. As usual in Continuum Mechanics we assume that there are two kinds of forces and couples acting on $\mathscr{P}$. The first ones are *body loads* acting on each material point of the body and which are independent of the rest of the body. The second ones are *contact loads* acting on the boundary of $\mathscr{P}$ and describing the interaction with the remaining part of the body or with the environment. In the case of micropolar continua one has the same situation as in general mechanics—we have to take into account forces and couples as primary variables.

The gravity force, centrifugal force, and ponderomotive force are examples of body forces. Body couples arise, for example, when the body is in an electromagnetic field.

Forces and couples acting on $\mathscr{P}$ can be represented as follows

$$\mathbf{f}(\mathscr{P}) = \mathbf{f}_B(\mathscr{P}) + \mathbf{f}_C(\mathscr{P}), \quad \mathbf{m}(\mathscr{P}) = \mathbf{m}_B(\mathscr{P}) + \mathbf{m}_C(\mathscr{P}),$$

where the subscript B denotes body forces and couples and C is for the surface quantities.

V. A. Eremeyev et al., *Foundations of Micropolar Mechanics*,
SpringerBriefs in Continuum Mechanics,
DOI: 10.1007/978-3-642-28353-6_3, © The Author(s) 2013

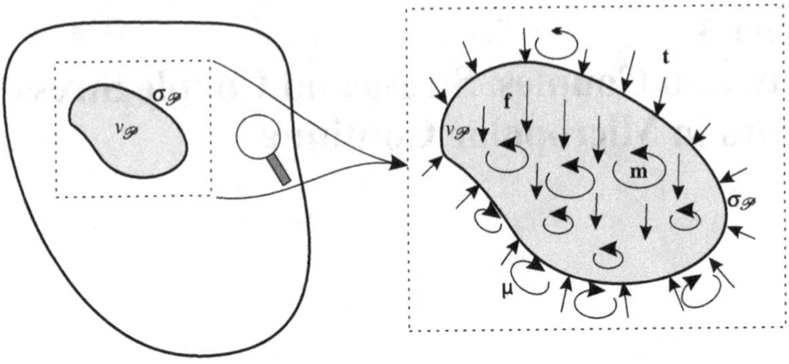

Fig. 3.1 Forces and couples acting on a body part $\mathscr{P}$

Let us suppose $\mathbf{f}_B(\mathscr{P})$, $\mathbf{m}_B(\mathscr{P})$ to be absolutely continuous functions of the mass of the part on which they act and $\mathbf{f}_C(\mathscr{P})$, $\mathbf{m}_C(\mathscr{P})$ to be absolutely continuous functions the surface area. This means that we can introduce their mass and surface densities:

$$\mathbf{f}_B(\mathscr{P}) = \iiint\limits_{v_{\mathscr{P}}} \rho \mathbf{f}\, \mathrm{d}v, \quad \mathbf{m}_B(\mathscr{P}) = \iiint\limits_{v_{\mathscr{P}}} \rho \mathbf{m}\, \mathrm{d}v,$$

$$\mathbf{f}_C(\mathscr{P}) = \iint\limits_{\sigma_{\mathscr{P}}} \mathbf{t}\, \mathrm{d}\sigma, \quad \mathbf{m}_C(\mathscr{P}) = \iint\limits_{\sigma_{\mathscr{P}}} \boldsymbol{\mu}\, \mathrm{d}\sigma,$$

where $v_{\mathscr{P}}$ is the volume of $\mathscr{P}$ in the actual configuration, $\sigma_{\mathscr{P}} = \partial v_{\mathscr{P}}$ is the boundary of $\mathscr{P}$, $\mathbf{t}$, $\boldsymbol{\mu}$ are the force and couple per the area unit in the actual configuration.

Definition 3.1 $\mathbf{t}$ is called the stress vector and $\boldsymbol{\mu}$ is the couple stress vector.

3.2 Euler's Laws of Motion

The angular velocity field $\boldsymbol{\omega}$ is independent of the linear velocity field $\mathbf{v}$ in a micropolar body. Let us introduce two definitions.

Definition 3.2 The momentum of part $\mathscr{P}$ of the body is

$$\mathfrak{P}(\mathscr{P}) \triangleq \iiint\limits_{v_{\mathscr{P}}} \rho \mathbf{v}\, \mathrm{d}v. \tag{3.1}$$

Definition 3.3 The moment of momentum of part $\mathscr{P}$ of the body is

$$\mathfrak{M}(\mathscr{P}) \triangleq \iiint\limits_{v_{\mathscr{P}}} \{(\mathbf{r} - \mathbf{r}_0) \times \rho \mathbf{v} + j\boldsymbol{\omega}\}\, \mathrm{d}v. \tag{3.2}$$

In the definitions the following variables are introduced: ρ is the material density, $\mathbf{r}$ is the position vector in the actual configuration, $\mathbf{r}_0$ is an arbitrary position vector that does not depend on t, j is the scalar measure of rotatory inertia of "microparticles" of the material.

Let us note that in the general case, j is a tensor so we have to change it to the tensor of inertia $\mathbf{J}$ that has to be positive definite. $\mathbf{J}$ is a characteristic of the material that defines the rotatory inertia of particles of the body. In Chap. 4 we discuss the generalizations of (3.1) and (3.2) that are the constitutive relations of special kind.

It should be noted the mentioned above definitions generalize the definitions of the momentum and the moment of momentum of the classical mechanics to continua, see Appendix B.

We accept the following two Eulerian dynamic laws as axioms.[1]

1 ***Balance of momentum. First Euler's law of motion.*** *The time rate of change of the momentum of an arbitrary part* $\mathscr{P}$ *of the body is equal to the total force acting on* $\mathscr{P}$:

$$\frac{\mathrm{d}}{\mathrm{d}t}\mathfrak{P}(\mathscr{P}) = \mathfrak{F}, \quad \mathfrak{F} \triangleq \iiint\limits_{v_{\mathscr{P}}} \rho \mathbf{f}\, \mathrm{d}v + \iint\limits_{\sigma_{\mathscr{P}}} \mathbf{t}\, \mathrm{d}\sigma. \tag{3.3}$$

2 ***Balance of moment of momentum. Second Euler's law of motion.*** *The time rate of change of the moment of momentum of an arbitrary part* $\mathscr{P}$ *of the body about a fixed point* $\mathbf{r}_0$ *is equal to the total moment about* $\mathbf{r}_0$ *acting on* $\mathscr{P}$:

$$\frac{\mathrm{d}}{\mathrm{d}t}\mathfrak{M}(\mathscr{P}) = \mathfrak{C},$$
$$\mathfrak{C} \triangleq \iiint\limits_{v_{\mathscr{P}}} \{(\mathbf{r} - \mathbf{r}_0) \times \rho \mathbf{f} + \rho \mathbf{m}\}\, \mathrm{d}v + \iint\limits_{\sigma_{\mathscr{P}}} \{(\mathbf{r} - \mathbf{r}_0) \times \mathbf{t} + \boldsymbol{\mu}\}\, \mathrm{d}\sigma. \tag{3.4}$$

Note that for an oriented medium, the momentum definition (3.1) and the formulation of the first law (3.3) coincide with the momentum definition for simple (non-polar) materials. For the medium with couple stresses, the definition of the moment of momentum (3.2) and the second law formulation (3.4) are different of those for simple materials; they contain additional terms related to the couple inertia tensor of a particle, the distributed body and surface couples.

Using the first dynamic law we can demonstrate that the formulation of the second law of motion does not depend on the choice of position vector $\mathbf{r}_0$.

[1] In what follows we consider an inertial reference frame.

3.3 Stress Tensor and Couple Stress Tensor

At instant t the stress vector and the couple stress tensor depend on the particle position $\mathbf{r}$ and the normal to surface $\mathbf{n}$ only.

Cauchy's postulate. *At a point of the body, the stress vector and the couple stress vector depend on the surface only through the unit normal to the considered surface, that is they have the same values, respectively, for all surfaces through the point which have the same normal.*

In what follows, the normal $\mathbf{n}$ to the body surface will be taken external with respect to the part of the body under consideration.

Cauchy's lemma. *The stress and couple stress vectors are odd functions with respect to* $\mathbf{n}$:

$$\mathbf{t}(\mathbf{r}, \mathbf{n}) = -\mathbf{t}(\mathbf{r}, -\mathbf{n}), \tag{3.5}$$

$$\boldsymbol{\mu}(\mathbf{r}, \mathbf{n}) = -\boldsymbol{\mu}(\mathbf{r}, -\mathbf{n}). \tag{3.6}$$

Cauchy's lemma is the re-formulation of the third Newton's axiom of reciprocal actions in micropolar continuum.

Equations (3.5) and (3.6) describe the interactions of the body parts that have contacting points.

Proof Consider an arbitrary part of the body occupying the domain $v_{\mathscr{P}}$. Let us take $\mathbf{r} \in v_{\mathscr{P}}$, cf. Fig. 3.2. The surface γ through $\mathbf{r}$ splits $v_{\mathscr{P}}$ into parts v_1 and v_2, $v_{\mathscr{P}} = v_1 \cup v_2$. Note that here v_1, v_2, and γ are also arbitrary. To prove relations (3.5) and (3.6) we use Euler's laws of motion for domains $v_{\mathscr{P}}$, v_1, and v_2. For a sufficiently smooth velocity field it is valid

$$\frac{\mathrm{d}}{\mathrm{d}t}\mathfrak{P}(\mathscr{P}) = \iiint\limits_{v_{\mathscr{P}}} \rho \frac{\mathrm{d}}{\mathrm{d}t}\mathbf{v}\,\mathrm{d}v.$$

Then the application of the first Euler's law (3.3) for $v_{\mathscr{P}}$ implies

$$\iiint\limits_{v_{\mathscr{P}}} \rho \frac{\mathrm{d}}{\mathrm{d}t}\mathbf{v}\,\mathrm{d}v = \iiint\limits_{v_{\mathscr{P}}} \rho \mathbf{f}\,\mathrm{d}v + \iint\limits_{\sigma_{\mathscr{P}}} \mathbf{t}\,\mathrm{d}\sigma. \tag{3.7}$$

Applying (3.3) on v_1 we get

$$\iiint\limits_{v_1} \rho \frac{\mathrm{d}}{\mathrm{d}t}\mathbf{v}\,\mathrm{d}v = \iiint\limits_{v_1} \rho \mathbf{f}\,\mathrm{d}v + \iint\limits_{\sigma_1} \mathbf{t}\,\mathrm{d}\sigma + \iint\limits_{\gamma} \mathbf{t}(\mathbf{n}_1)\,\mathrm{d}\sigma, \tag{3.8}$$

where σ_1 is a part of σ that belongs to v_1 and $\mathbf{n}_1$ is the external unit normal to γ.

Similarly, for v_2 we have

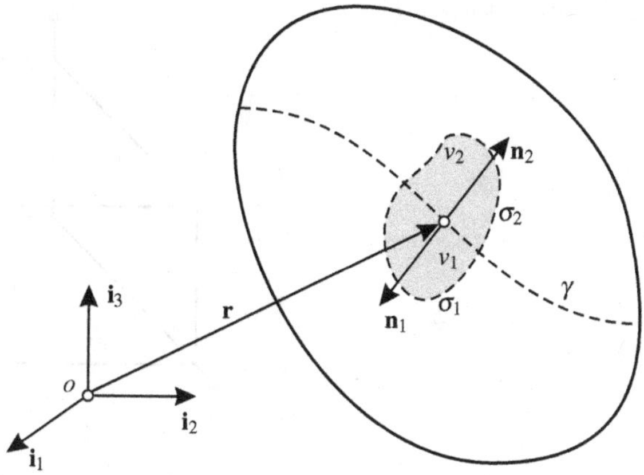

Fig. 3.2 Arbitrary part $\mathscr{P}$ of the body $\mathscr{B}$

$$\iiint\limits_{v_2} \rho \frac{\mathrm{d}}{\mathrm{d}t}\mathbf{v}\,\mathrm{d}v = \iiint\limits_{v_2} \rho\mathbf{f}\,\mathrm{d}v + \iint\limits_{\sigma_2} \mathbf{t}\,\mathrm{d}\sigma + \iint\limits_{\gamma} \mathbf{t}(\mathbf{n}_2)\,\mathrm{d}\sigma. \tag{3.9}$$

Subtracting (3.8), (3.9) from equality (3.7) we get

$$\mathbf{0} = \iint\limits_{\gamma} \mathbf{t}(\mathbf{n}_1)\,\mathrm{d}\sigma + \iint\limits_{\gamma} \mathbf{t}(\mathbf{n}_2)\,\mathrm{d}\sigma. \tag{3.10}$$

Supposing the integrands in (3.10) to be sufficiently smooth we derive

$$\mathbf{t}(\mathbf{n}_1) + \mathbf{t}(\mathbf{n}_2) = \mathbf{0}.$$

As $\mathbf{n}_1 = -\mathbf{n}_2$, it follows the equality (3.5) that completes the proof of the first part of the lemma.

The proof of equality (3.6) is similar. It holds

$$\frac{\mathrm{d}}{\mathrm{d}t}\mathfrak{M}(\mathscr{P}) = \iiint\limits_{v_{\mathscr{P}}} \left\{(\mathbf{r}-\mathbf{r}_0)\times\rho\frac{\mathrm{d}}{\mathrm{d}t}\mathbf{v} + j\frac{\mathrm{d}}{\mathrm{d}t}\boldsymbol{\omega}\right\}\mathrm{d}v.$$

Applying second dynamic law (3.4) for domains $v_{\mathscr{P}}$, v_1, v_2 and repeating the transformations we get

$$\mathbf{0} = \iint\limits_{\gamma} \{(\mathbf{r}-\mathbf{r}_0)\times(\mathbf{t}(\mathbf{n}_1)+\mathbf{t}(\mathbf{n}_2)) + \boldsymbol{\mu}(\mathbf{n}_1)+\boldsymbol{\mu}(\mathbf{n}_2)\}\,\mathrm{d}\sigma. \tag{3.11}$$

Using (3.5) from (3.11) it follows the second equality (3.6) that completes the proof. □

Fig. 3.3 Parallelepiped Π

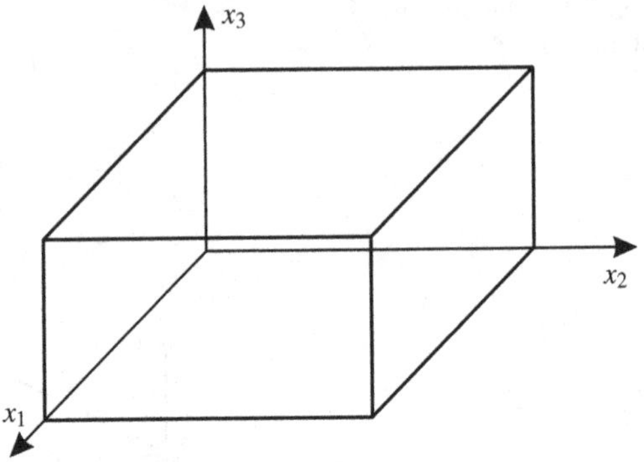

Cauchy's lemma is an essential tool for the introduction of the stress tensor and the couple stress tensor that are presented by *Cauchy's theorem.*

Theorem 3.1 *For any point of the body there exist second-order tensors* $\mathbf{T}$ *and* $\mathbf{M}$ *such that the stress vector and the couple stress vector acting at a point of the surface with normal* $\mathbf{n}$ *are presented by formulae*

$$\mathbf{t} = \mathbf{T} \cdot \mathbf{n}, \quad \boldsymbol{\mu} = \mathbf{M} \cdot \mathbf{n}.$$

Proof First let us prove the existence of $\mathbf{T}$. Consider an arbitrary orthogonal parallelepiped Π with edges oriented along the axes of Cartesian coordinates x_1, x_2, x_3, cf. Fig. 3.3. The unit normal to a side of the parallelepiped coincides with one of the coordinate unit vectors $\mathbf{n} = \pm\mathbf{i}_k$ $(k = 1, 2, 3)$ up to direction. In frame $\mathbf{i}_k$, the stress vector is

$$\mathbf{t}(\mathbf{r}, \mathbf{H}, \mathbf{i}_k) = t_{sk}(\mathbf{r}, \mathbf{H})\mathbf{i}_s. \tag{3.12}$$

In the rest part of the proof we will omit arguments $\mathbf{r}$ and $\mathbf{H}$. Here t_{sk} are the components of $\mathbf{t}$ in the frame $\mathbf{i}_k$. From the law of reciprocal actions it follows

$$\mathbf{t}(-\mathbf{i}_k) = -t_{sk}\mathbf{i}_s.$$

We denote by n_k the components of the normal in the frame $\mathbf{i}_k$. For coordinate area elements, the representation (3.12) is

$$\mathbf{t}(\mathbf{i}_1) = n_1 t_{s1}\mathbf{i}_s, \quad \mathbf{t}(\mathbf{i}_2) = n_2 t_{s2}\mathbf{i}_s, \quad \mathbf{t}(\mathbf{i}_3) = n_3 t_{s3}\mathbf{i}_s. \tag{3.13}$$

Let us apply the first Euler's law to the parallelepiped. We get

$$\iiint\limits_{v_\Pi} \rho\left(\frac{\mathrm{d}\mathbf{v}}{\mathrm{d}t} - \mathbf{f}\right) \mathrm{d}v = \iint\limits_{\sigma_\Pi} \mathbf{t}\,\mathrm{d}\sigma.$$

Using (3.13) the latter equation can be written as

$$\iiint\limits_{v_\Pi} \rho\left(\frac{\mathrm{d}\mathbf{v}}{\mathrm{d}t} - \mathbf{f}\right) \mathrm{d}v = \iint\limits_{\sigma_\Pi} n_k t_{sk} \mathbf{i}_s \, \mathrm{d}\sigma.$$

Applying the Gauss–Ostrogradsky theorem to the surface integral we get

$$\iiint\limits_{v_\Pi} \left\{\rho\left(\frac{\mathrm{d}\mathbf{v}}{\mathrm{d}t} - \mathbf{f}\right) - \frac{\partial t_{sk}}{\partial x_k}\mathbf{i}_s\right\} \mathrm{d}v = \mathbf{0}. \tag{3.14}$$

As the parallelepiped is arbitrary, from the integral equation (3.14) it follows the differential equation

$$\rho\left(\frac{\mathrm{d}\mathbf{v}}{\mathrm{d}t} - \mathbf{f}\right) - \frac{\partial t_{sk}}{\partial x_k}\mathbf{i}_s = \mathbf{0}. \tag{3.15}$$

Let us consider now an arbitrary tetrahedron T, Fig. 3.4. Applying the first Euler's law for T we get

$$\iiint\limits_{v_T} \rho\left(\frac{\mathrm{d}\mathbf{v}}{\mathrm{d}t} - \mathbf{f}\right) \mathrm{d}v = \iint\limits_{\sigma_T} n_k t_{sk} \mathbf{i}_s \, \mathrm{d}\sigma + \iint\limits_{M_1 M_2 M_3} \mathbf{t}(\mathbf{n}) \, \mathrm{d}\sigma, \tag{3.16}$$

where v_T is the volume, σ_T is the part of the tetrahedron sides that are parallel to the coordinate planes, $M_1 M_2 M_3$ is the inclined side. In (3.16) we have used the representation of the stress vector on the coordinate plane elements (3.13).

With the help of (3.15) we transform the left-hand side of (3.16)

$$\iiint\limits_{v_T} \frac{\partial t_{sk}}{\partial x_k}\mathbf{i}_s \, \mathrm{d}v = \iint\limits_{\sigma_T} n_k t_{sk} \mathbf{i}_s \, \mathrm{d}\sigma + \iint\limits_{M_1 M_2 M_3} \mathbf{t}(\mathbf{n}) \, \mathrm{d}\sigma. \tag{3.17}$$

Applying again the Gauss–Ostrogradsky formula to the volume integral in (3.17) we transform the equation to the form

$$\iint\limits_{M_1 M_2 M_3} [\mathbf{t}(\mathbf{n}) - n_k t_{sk} \mathbf{i}_s] \, \mathrm{d}\sigma = \mathbf{0}.$$

As the tetrahedron and its side $M_1 M_2 M_3$ are arbitrary it follows the identity

$$\mathbf{t}(\mathbf{n}) - n_k t_{sk} \mathbf{i}_s = \mathbf{0}$$

that is valid for any normal $\mathbf{n}$. Thus, we have demonstrated that $\mathbf{t}$ depends on $\mathbf{n}$ linearly.

Fig. 3.4 Tetrahedron T

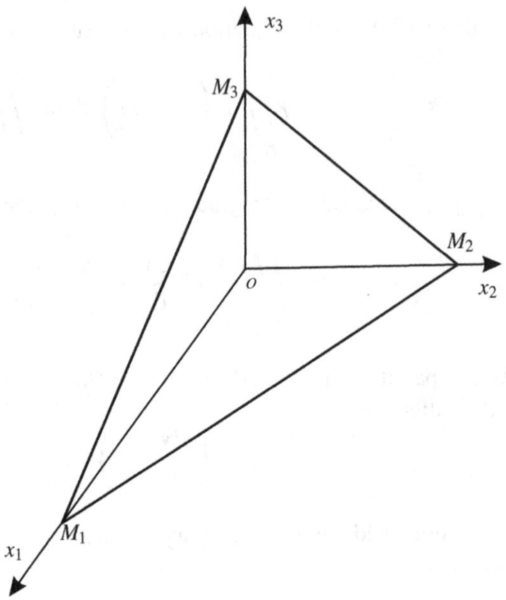

It is well known the following representation theorem of a linear function, see Theorem A.2 or [4].

Theorem 3.2 *A linear vector-valued function* $\mathbf{l}(\mathbf{n})$ *of a vectorial argument* $\mathbf{n}$ *can be represented in the form* $\mathbf{l}(\mathbf{n}) = \mathsf{L} \cdot \mathbf{n}$, *where* L *is a second-order tensor.*

By this theorem, there exists T such that

$$\mathbf{t}(\mathbf{n}) = \mathsf{T} \cdot \mathbf{n}. \tag{3.18}$$

Thus the first part of Cauchy's theorem is proved.

The second theorem part on the existence of M can be proven similarly. First we should apply the second Euler's law first to an arbitrary parallelepiped and then to an arbitrary tetrahedron.

Let us represent the couple stress vector $\boldsymbol{\mu}$ on the coordinate plane elements as follows

$$\boldsymbol{\mu}(\mathbf{i}_k) = m_{sk}\mathbf{i}_s.$$

With the use of the components of the unit normal on the coordinate plane elements it can be written as

$$\boldsymbol{\mu}(\mathbf{i}_1) = n_1 m_{s1}\mathbf{i}_s, \quad \boldsymbol{\mu}(\mathbf{i}_2) = n_2 m_{s2}\mathbf{i}_s, \quad \boldsymbol{\mu}(\mathbf{i}_3) = n_3 m_{s3}\mathbf{i}_s. \tag{3.19}$$

Here we have used the equality $\boldsymbol{\mu}(-\mathbf{i}_k) = -m_{sk}\mathbf{i}_s$.

Applying the second Euler's law to Π we get

$$\iiint\limits_{v_\Pi} \left\{ (\mathbf{r} - \mathbf{r}_0) \times \rho \left(\frac{\mathrm{d}\mathbf{v}}{\mathrm{d}t} - \mathbf{f} \right) + j \frac{\mathrm{d}\boldsymbol{\omega}}{\mathrm{d}t} - \rho \mathbf{m} \right\} \mathrm{d}v$$
$$= \iint\limits_{\sigma_\Pi} \{ (\mathbf{r} - \mathbf{r}_0) \times \mathbf{t} + \boldsymbol{\mu} \} \, \mathrm{d}\sigma. \tag{3.20}$$

Using (3.13) and (3.19), we rewrite the surface integral in (3.20) as follows

$$\iint\limits_{\sigma_\Pi} \{ (\mathbf{r} - \mathbf{r}_0) \times n_k t_{sk} \mathbf{i}_s + n_k m_{sk} \mathbf{i}_s \} \, \mathrm{d}\sigma.$$

Applying the Gauss–Ostrogradsky theorem, we transform the integral over σ_Π to the integral over volume v_Π

$$\iiint\limits_{v_\Pi} \frac{\partial}{\partial x_k} \{ (\mathbf{r} - \mathbf{r}_0) \times t_{sk} \mathbf{i}_s + m_{sk} \mathbf{i}_s \} \, \mathrm{d}v.$$

Thus with regard to identity $\partial \mathbf{r} / \partial x_k = \mathbf{i}_k$ we reduce Eq. (3.20) to the form

$$\iiint\limits_{v_\Pi} \left\{ (\mathbf{r} - \mathbf{r}_0) \times \left[\rho \left(\frac{\mathrm{d}\mathbf{v}}{\mathrm{d}t} - \mathbf{f} \right) - \frac{\partial t_{sk}}{\partial x_k} \mathbf{i}_s \right] \right.$$
$$\left. + j \frac{\mathrm{d}\boldsymbol{\omega}}{\mathrm{d}t} - \rho \mathbf{m} - t_{sk} \mathbf{i}_k \times \mathbf{i}_s - \frac{\partial m_{sk}}{\partial x_k} \mathbf{i}_s \right\} \mathrm{d}v = \mathbf{0}.$$

Using the first Euler's law we have shown that the expression in square brackets is equal to zero. So we get the equation

$$\iiint\limits_{v_\Pi} \left\{ j \frac{\mathrm{d}\boldsymbol{\omega}}{\mathrm{d}t} - \rho \mathbf{m} - t_{sk} \mathbf{i}_k \times \mathbf{i}_s - \frac{\partial m_{sk}}{\partial x_k} \mathbf{i}_s \right\} \mathrm{d}v = \mathbf{0},$$

from which it follows the differential equation

$$j \frac{\mathrm{d}\boldsymbol{\omega}}{\mathrm{d}t} - \rho \mathbf{m} = \frac{\partial m_{sk}}{\partial x_k} \mathbf{i}_s - t_{sk} \mathbf{i}_s \times \mathbf{i}_k. \tag{3.21}$$

The last addendum in (3.21) is *the vector invariant* of the second-order tensor T defined by Eq. (A.4),

$$\mathsf{T}_\times \triangleq t_{ks} \mathbf{i}_k \times \mathbf{i}_s.$$

Next, we will show the representation (3.19) is valid for any area element. Applying the second Euler's law to an arbitrary tetrahedron we get

$$\iiint\limits_{v_T} \left\{(\mathbf{r}-\mathbf{r}_0)\times\rho\left(\frac{d\mathbf{v}}{dt}-\mathbf{f}\right)+j\frac{d\boldsymbol{\omega}}{dt}-\rho\mathbf{m}\right\}dv$$
$$=\iint\limits_{\sigma_T}\{(\mathbf{r}-\mathbf{r}_0)\times\mathbf{t}+n_k m_{sk}\mathbf{i}_s\}\,d\sigma+\iint\limits_{M_1M_2M_3}\{(\mathbf{r}-\mathbf{r}_0)\times\mathbf{t}+\boldsymbol{\mu}(\mathbf{N})\}\,d\sigma. \tag{3.22}$$

Let us reduce (3.22) to an equation that contains only a surface integral over side $M_1M_2M_3$. For this, in (3.22) we transform the terms having factor $(\mathbf{r}-\mathbf{r}_0)$. With regard of (3.18) and the identity

$$\operatorname{div}\left[(\mathbf{r}-\mathbf{r}_0)\times\mathbf{T}\right]=(\mathbf{r}-\mathbf{r}_0)\times\operatorname{div}\mathbf{T}-\mathbf{T}_\times$$

we can show there holds the equality

$$\iiint\limits_{v_T}\left\{(\mathbf{r}-\mathbf{r}_0)\times\rho\left(\frac{d\mathbf{v}}{dt}-\mathbf{f}\right)\right\}dv-\iint\limits_{\partial v_T}\{(\mathbf{r}-\mathbf{r}_0)\times\mathbf{t}\}\,d\sigma=-\iiint\limits_{v_T}\mathbf{T}_\times\,dv,$$

where $\partial v_T=\sigma_T\bigcup M_1M_2M_3$. Thus the volume integral in (3.22) takes the form

$$\iiint\limits_{v_T}\left(j\frac{d\boldsymbol{\omega}}{dt}-\rho\mathbf{m}+\mathbf{T}_\times\right)dv.$$

With use of (3.21) we transform (3.22) to the necessary form

$$\iint\limits_{M_1M_2M_3}\{\boldsymbol{\mu}(\mathbf{n})-n_k m_{sk}\mathbf{i}_s(\mathbf{n})\}\,d\sigma=\mathbf{0}.$$

As the tetrahedron is arbitrary, this shows function $\boldsymbol{\mu}(\mathbf{n})$ is linear with respect to $\mathbf{n}$ and so there holds the representation

$$\boldsymbol{\mu}(\mathbf{n})=\mathbf{M}\cdot\mathbf{n}. \tag{3.23}$$

Equation (3.23) completes the proof of Cauchy's theorem for the Cosserat medium. □

Definition 3.4 Tensor $\mathbf{T}$ is called the stress tensor of Cauchy-type and tensor $\mathbf{M}$ the couple stress tensor of Cauchy-type.

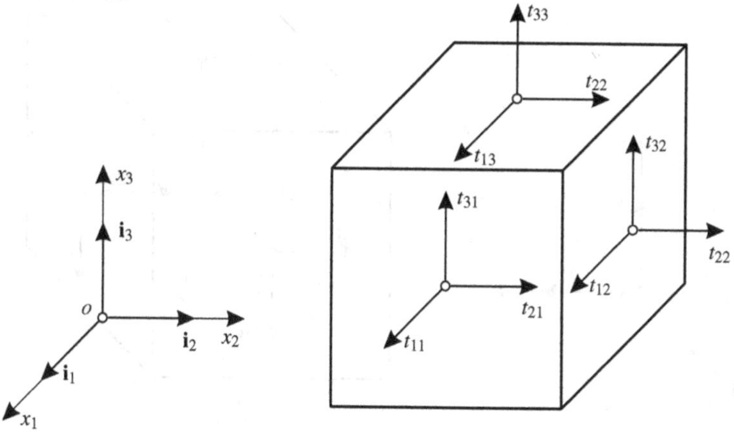

Fig. 3.5 Positive stresses

So from the theorem proof we see that matrices t_{sk} and m_{sk} represent the components of the stress tensor and of the couple stress tensor in Cartesian basis $\mathbf{i}_k$:

$$\mathbf{T} = t_{ks}\mathbf{i}_k\mathbf{i}_s, \quad \mathbf{M} = m_{ks}\mathbf{i}_k\mathbf{i}_s.$$

It should be noted an important property of the stress and couple stress tensors: they are not symmetric, in general, that is $\mathbf{T}^T \neq \mathbf{T}$, $\mathbf{M}^T \neq \mathbf{M}$. This property distinguishes the micropolar continuum from the simple materials for which the stress tensor is always symmetric.

Non-symmetry of matrices t_{sk} and m_{sk} makes us carefully work with their indices. Let us consider an arbitrary cube in the body. The sides of the cube are oriented along the axes of Cartesian coordinate system. Tangent and normal stresses that act on the surface cube are shown on Fig. 3.5. Second subscript of t_{sk} means the area element with normal $\mathbf{i}_k$ whereas the first subscript shows the direction of the stress action that is $\mathbf{i}_s$. For example, t_{13} is the stress acting on the cross-section of the body that is perpendicular to axis x_3, its action is in the direction of axis x_1. Note this agreement for the indices is valid for any orthonormal basis.

To formulate static boundary conditions, we hold the following rule for the choice of sign for the components of the stress tensor.

The rule of signs for stress tensor. *Normal stresses components are positive if they are stretching, they are negative being compressive. A tangent stress component is positive if it acts on a plane element with the normal that coincides with the basis vector and it is co-directed with one of the basis vectors. A tangent stress component that acts on a plane element with the normal that is opposite to a basis vector, is positive if it is opposite to another basis vector.*

By this rule, the stresses depicted on Fig. 3.5 are positive. On Fig. 3.6 we show negative stresses that act on the sides of cube that are seen.

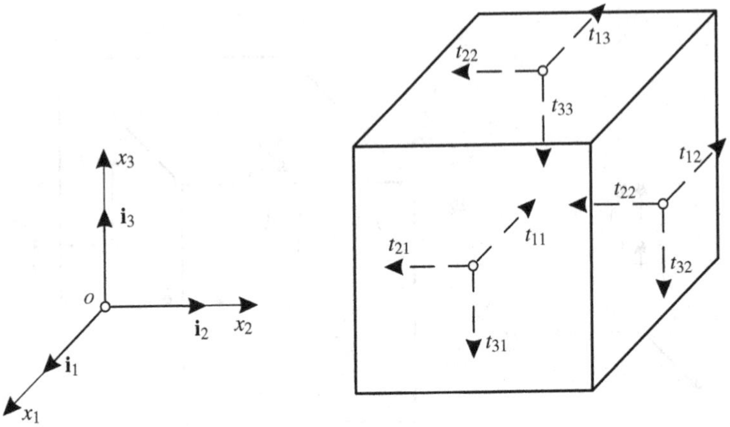

Fig. 3.6 Negative stresses

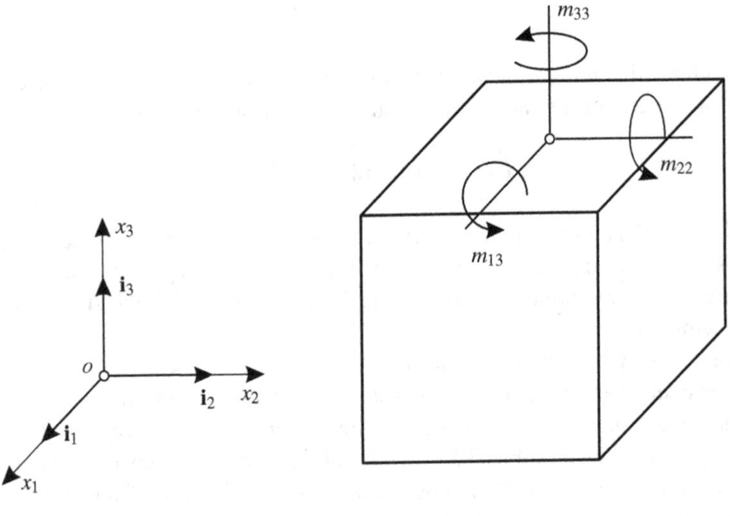

Fig. 3.7 Couple stresses

Everything we have said above for the stress tensor components is valid for the couple stress tensor as well. So the second subscript of m_{sk} denotes the plane element with normal $\pm\mathbf{i}_k$ whereas first subscript defines the direction of action, that is $\mathbf{i}_s$. It is traditional to depict the couples and couple stresses with rounded arrows as any axial vector, see Sect. A.6. For example, Fig. 3.7 represents couple stresses that act on a plane element with the normal $\mathbf{i}_3$. It is seen the normal couple stresses represent the torsion torque whereas the tangent couple stresses are for the bending couples.

However, for the notation of stress vectors and couple stress vectors it is more convenient their depicting with direct arrows. To differ the vectors that describe the couple stresses from the stress vectors, we will depict them by double arrows.

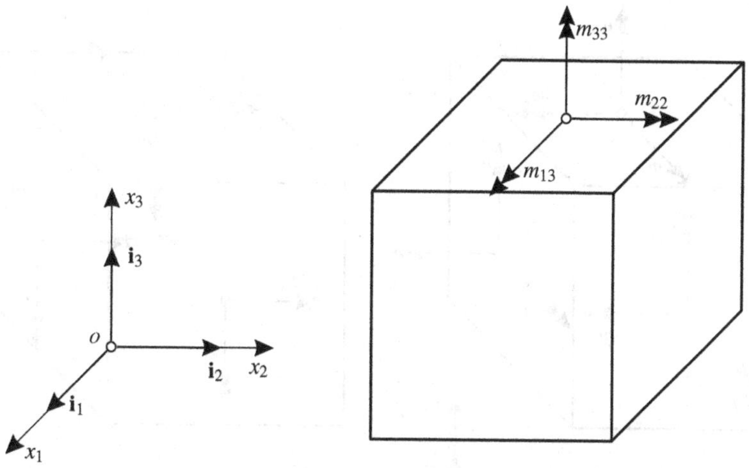

Fig. 3.8 Another notation for couple stresses

Figure 3.8 uses this picture representation for the couple stress vectors of Fig. 3.7. A couple that corresponds to a double arrow acts in the clockwise direction if to see along the arrow direction.

The sign rule for the couple stresses is completely analogous to the sign rule for the stresses up to the notation. On Fig. 3.9, positive couple stresses are shown with continuum double arrows whereas the negative ones are depicted with dotted arrows.

Let us formulate the sign rule for the couple stresses.

The sign rule for the couple stresses. *Distributed on a plane element a torsion torque is positive when it acts in clockwise direction if to see along the external to the area element normal.*

Let a bending couple stress, a tangent couple stress, act on an area element with the normal that is co-directed with a basis vector. It is positive if it acts in the counterclockwise direction when we see in the direction that is opposite to another basis vector. For a bending couple stress on an area element with the normal that is opposite to a basis vector, the positive direction is if see the clockwise action of the couple stress when seeing in the direction that is opposite to another basis vector.

3.4 Principal Stresses in Micropolar Continua

The representation of the Cauchy stress tensor in a general non-orthogonal basis is

$$\mathsf{T} = t_{ks}\mathbf{i}_s \otimes \mathbf{i}_k,$$

where matrix t_{sk} has nine components that are non-zero, in general.

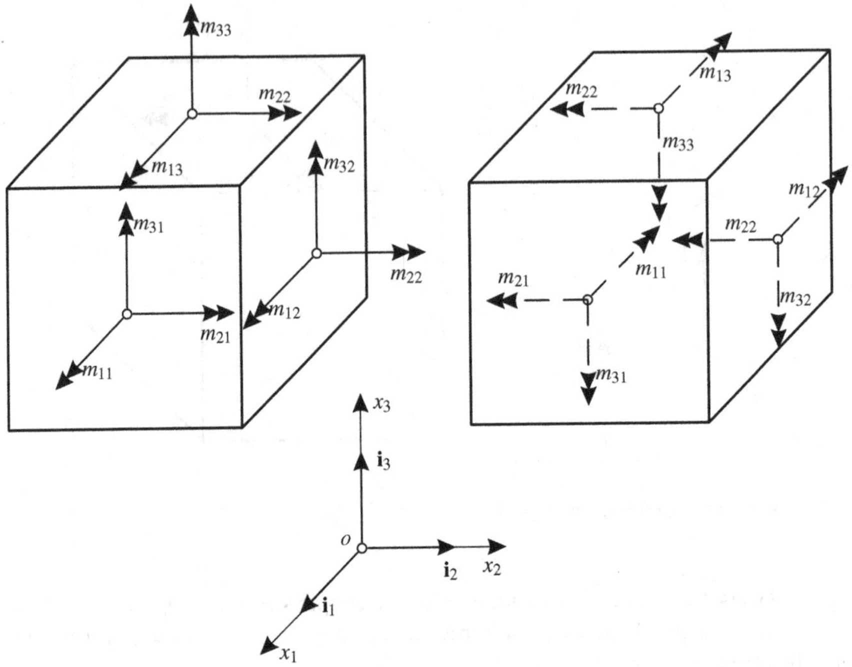

Fig. 3.9 Positive and negative couple stresses

Let us consider the problem of finding a basis in which matrix representation of $\mathbf{T}$ is most simple.

It is well known that for the classical Cauchy continuum with an symmetric stress tensor, the matrix of representation is diagonal

$$\mathbf{T} = \sigma_1 \mathbf{e}_1 \otimes \mathbf{e}_1 + \sigma_2 \mathbf{e}_2 \otimes \mathbf{e}_2 + \sigma_3 \mathbf{e}_3 \otimes \mathbf{e}_3, \tag{3.24}$$

where $\sigma_1, \sigma_2, \sigma_3$ are eigenvalues of the matrix

$$\begin{pmatrix} t_{11} & t_{12} & t_{13} \\ t_{21} & t_{22} & t_{23} \\ t_{31} & t_{32} & t_{33} \end{pmatrix},$$

that are called *principal stresses*, and $\mathbf{e}_1, \mathbf{e}_2, \mathbf{e}_3$, are eigenvectors of t_{ks} that are called *principal axes* of $\mathbf{T}$. The principle axes are the normals to the surface elements on which tangential stresses are vanish. As for the Cauchy continuum $\mathbf{T}$ is always symmetric, there exists an orthonormal basis $\mathbf{e}_1$, $\mathbf{e}_2$, $\mathbf{e}_3$ in which the matrix for $\mathbf{T}$ is diagonal. Equation (3.24) is called *spectral decomposition* of second-order symmetric tensor $\mathbf{T}$. From the physical point of view the spectral decomposition (3.24) means that at any point of the body we can always find orthogonal surface elements on which only normal stresses, stretching or compressing, act. If the stress field is homogeneous then we can select such an elementary cube that on its facets there act normal stresses only.

Unfortunately, in micropolar mechanics, $\mathbf{T}$ is not symmetric, in general, and so a spectral decomposition of $\mathbf{T}$ can not be established. Now an analogue of the orthogonal representation (3.24) of $\mathbf{T}$ is the matrix singular value decomposition [5] that exists for all non-symmetric tensors.

Definition 3.5 We call the singular value decomposition of a second-order tensor $\mathbf{T}$ the following relation

$$\mathbf{T} = s_1 \mathbf{e}_1 \otimes \mathbf{e}_1' + s_2 \mathbf{e}_2 \otimes \mathbf{e}_2' + s_3 \mathbf{e}_3 \otimes \mathbf{e}_3', \tag{3.25}$$

where s_k $(k = 1, 2, 3)$ are non-negative real numbers, that are singular values of $\mathbf{T}$, and $\mathbf{e}_k$, $\mathbf{e}_j'$ are two orthonormal bases.

Proof Let us prove that the singular value decomposition of an arbitrary $\mathbf{T}$ exists. It follows from the polar decomposition of $\mathbf{T} = \mathbf{S} \cdot \mathbf{Q}$, cf. [5], where $\mathbf{S} = (\mathbf{T} \cdot \mathbf{T}^T)^{1/2}$ is a non-strictly positive definite symmetric tensor and $\mathbf{Q}$ is an orthogonal tensor. Indeed let us denote eigenvalues of $\mathbf{S}$ by s_k. Let $\mathbf{e}_k$ be an orthonormal basis of the eigenvectors of $\mathbf{S}$. Let $\mathbf{e}_k' = \mathbf{e}_k \cdot \mathbf{Q}$. The latter is an orthonormal basis as well. In the polar decomposition of $\mathbf{T}$, for $\mathbf{S}$ we substitute its orthogonal representation

$$\mathbf{S} = s_1 \mathbf{e}_1 \otimes \mathbf{e}_1 + s_2 \mathbf{e}_2 \otimes \mathbf{e}_2 + s_3 \mathbf{e}_3 \otimes \mathbf{e}_3.$$

Then it follows the representation (3.25). □

The singular representation (3.25) includes two orthogonal bases that is its disadvantage.

Let us answer the question whether there exist a more simple representation of non-symmetric $\mathbf{T}$ with only one basis that can be non-orthonormal. We will use the results on the *real Jordan canonical form* of a non-symmetric real matrix, cf. [5]. It is shown the following. A non-symmetric real valued 3×3 matrix A is similar to one of the following matrices

$$\begin{pmatrix} \lambda_1 & 0 & 0 \\ 0 & \lambda_2 & 0 \\ 0 & 0 & \lambda_3 \end{pmatrix}, \tag{3.26}$$

$$\begin{pmatrix} \lambda_1 & 0 & 0 \\ 0 & \lambda_2 & 0 \\ 0 & 0 & \lambda_2 \end{pmatrix}, \quad \begin{pmatrix} \lambda_1 & 0 & 0 \\ 0 & \lambda_2 & \varepsilon \\ 0 & 0 & \lambda_2 \end{pmatrix}, \tag{3.27}$$

$$\begin{pmatrix} \lambda_1 & 0 & 0 \\ 0 & \lambda_1 & 0 \\ 0 & 0 & \lambda_1 \end{pmatrix}, \quad \begin{pmatrix} \lambda_1 & 0 & 0 \\ 0 & \lambda_1 & \varepsilon \\ 0 & 0 & \lambda_1 \end{pmatrix}, \quad \begin{pmatrix} \lambda_1 & \varepsilon & 0 \\ 0 & \lambda_1 & \varepsilon \\ 0 & 0 & \lambda_1 \end{pmatrix}, \tag{3.28}$$

$$\begin{pmatrix} \lambda_1 & 0 & 0 \\ 0 & \alpha & \beta \\ 0 & -\beta & \alpha \end{pmatrix} \tag{3.29}$$

with real valued similarity matrices. Let us recall that two matrices A and B are called similar if $A = P^{-1}BP$ for some invertible matrix P. Which of the matrices should be selected, it depends on eigenvalues of A and their multiplicity. Here λ_k are real valued eigenvalues of A, ε is an arbitrarily small non-zero number, α and β are real and imaginary parts of a complex eigenvalue of A when its exists. Let us note that usually one takes $\varepsilon = 1$, cf. [5]. This corresponds to a selection of the similarity matrix. The case (3.26) is for existence of three different real eigenvalues $\lambda_1 \neq \lambda_2 \neq \lambda_3$ of A. The case (3.27) happens when $\lambda_1 \neq \lambda_2 = \lambda_3$ and they are real. If $\lambda_1 = \lambda_2 = \lambda_3$ then the representation (3.28) is valid. The fourth case (3.29) holds if one of eigenvalues is real and other two of eigenvalues are complex, $\lambda_1 \neq \lambda_2 = \overline{\lambda_3} \equiv \alpha + i\beta$, and $\overline{(\ldots)}$ denotes the complex conjugate of $(\ldots)$.

Thus, a non-symmetric $\mathbf{T}$ has a representation

$$\mathbf{T} = t^{\circ}_{mn}\mathbf{e}_m \otimes \mathbf{e}_n,$$

where t°_{mn} is one of the matrices given by Eqs. (3.26)–(3.29). Here vectors $\mathbf{e}_m$ are not orthogonal, in general.

Obviously, all these results on the representation of $\mathbf{T}$ relate to the couple stress tensor $\mathbf{M}$, that is $\mathbf{M}$ takes form $\mathbf{M} = m^{\circ}_{mn}\tilde{\mathbf{e}}_m \otimes \tilde{\mathbf{e}}_n$, where m°_{mn} is the matrix having the structure of one of (3.26)–(3.29) and $\tilde{\mathbf{e}}_n$ is a non-orthogonal basis.

3.5 Equations of Motion

Transforming Euler's laws of motion as was done in the proof of Cauchy's theorem 3.1, we get dynamic equations of micropolar continuum in the local form:

$$\rho \frac{d\mathbf{v}}{dt} = \operatorname{div} \mathbf{T} + \rho \mathbf{f}, \tag{3.30}$$

$$j \frac{d\boldsymbol{\omega}}{dt} = \operatorname{div} \mathbf{M} - \mathbf{T}_{\times} + \rho \mathbf{m}. \tag{3.31}$$

Deriving Eq. (3.31) we have used Eq. (3.30).

Equations (3.30) and (3.31) in Cartesian coordinates take the form

$$\rho \frac{dv_s}{dt} = \frac{\partial t_{sk}}{\partial x_k} + \rho f_s, \quad j \frac{d\omega_s}{dt} = \frac{\partial m_{sk}}{\partial x_k} + t_{mn}\varepsilon_{mns} + \rho m_s, \tag{3.32}$$

where $\varepsilon_{mns} = -(\mathbf{i}_m \times \mathbf{i}_n) \cdot \mathbf{i}_s$.

When the medium does not possesses couple's properties, that is rotation interaction of particles is negligible, then in Eq. (3.31) we should change to zero the following terms: the rotation inertia j, the couple stress tensor $\mathbf{M}$ and the volume couples $\mathbf{m}$. As a consequence of the balance of moment of momentum we obtain the following equation:

$$\mathbf{T}_\times = \mathbf{0}. \tag{3.33}$$

Its solution is the symmetric stress tensor, that is $\mathbf{T} = \mathbf{T}^T$. Thus when couple stresses and the distributed external couples in the balance equation of moment of momentum are absent, it follows the symmetry of the Cauchy stress tensor that is a property of the classic continuum mechanics. However, now it is impossible to consider the action of the couple loads, in particular the action of the couples distributed on the boundary or inside the body.

For classical continuum, that is when couple stresses and the external couples are absent, Eq. (3.33) holds automatically as a consequence of the constitutive equations. Hence, the balance equation of moment of momentum plays less important role in the Cauchy continuum.

3.6 Boundary-Value Problems

To setup a boundary-value problem for a micropolar body, we should supplement the motion equations (3.30) and (3.31) with boundary and initial conditions. *Static* boundary conditions consist of external forces and couples $\mathbf{t}_*$ and $\boldsymbol{\mu}_*$ given on the body boundary or on some its part $\sigma_f \subset \sigma$ (in the actual configuration):

$$\mathbf{T} \cdot \mathbf{n} = \mathbf{t}_*, \quad \mathbf{M} \cdot \mathbf{n} = \boldsymbol{\mu}_* \quad \text{on} \quad \sigma_f \tag{3.34}$$

as is shown on Fig. 3.10.

On the rest part of the boundary we assign two *kinematic* conditions

$$\mathbf{r} = \mathbf{r}_*, \quad \mathbf{H} = \mathbf{H}_* \quad (\mathbf{H}_* \cdot \mathbf{H}_*^T = \mathbf{I}) \quad \text{on} \quad \sigma_u, \tag{3.35}$$

where there are given $\mathbf{r}_*$ and $\mathbf{H}_*$, the translations and microrotations of the body particles on $\sigma_u = \sigma/\sigma_f$. We get a particular case of kinematic boundary conditions, when the part σ_u is fixed, that is

$$\mathbf{r} = \mathbf{R}, \quad \mathbf{H} = \mathbf{I} \quad \text{on} \quad \sigma_u. \tag{3.36}$$

Using the vectorial representation of $\mathbf{H}$ through the finite rotation vector $\boldsymbol{\theta}$, see (A.8), one can rewrite the kinematical boundary conditions:

$$\mathbf{u} = \mathbf{u}_*, \quad \boldsymbol{\theta} = \boldsymbol{\theta}_* \quad \text{on} \quad \sigma_u \tag{3.37}$$

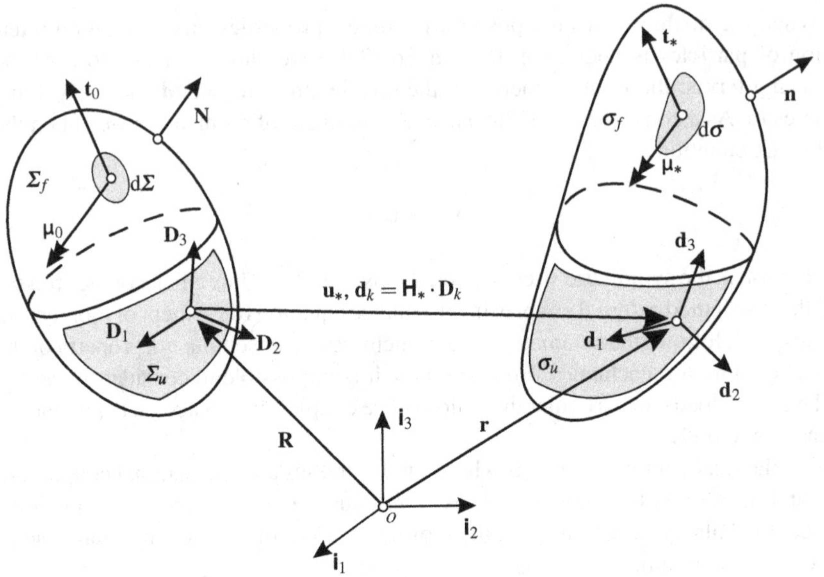

Fig. 3.10 Boundary conditions. On the *left* the body is shown in the reference configuration, on the *right* it is in the actual configuration

with given $\mathbf{u}_*$ and $\boldsymbol{\theta}_*$. In particular, Eq. (3.36) take simple form: $\mathbf{u} = \mathbf{0}, \boldsymbol{\theta} = \mathbf{0}$ on σ_u.

There are other types of boundary conditions of the micropolar mechanics. For example, in the micropolar hydrodynamics on a free surface they use the boundary condition: $\boldsymbol{\omega} = -\frac{1}{2}\operatorname{rot}\mathbf{v}$, see [6].

For dynamic problems we should assign the initial conditions

$$\mathbf{r} = \mathbf{r}^0, \quad \mathbf{v} = \mathbf{v}^0, \qquad \mathbf{H} = \mathbf{H}^0 \ \ (\mathbf{H}^0 \cdot \mathbf{H}^{0^T} = \mathbf{I}), \quad \boldsymbol{\omega} = \boldsymbol{\omega}^0 \ \text{ at } \ t = t_0 \qquad (3.38)$$

with given $\mathbf{r}^0$, $\mathbf{v}^0$, $\mathbf{H}^0$, and $\boldsymbol{\omega}^0$.

Equations (3.30) and (3.31) with boundary conditions (3.34), (3.35) or (3.36) and with initial conditions (3.38) constitute the Eulerian boundary-value problem for the micropolar body, which is widely used in the micropolar hydrodynamics.

Using *Piola's identity*

$$\operatorname{Div}\left[(\det \mathbf{F})\mathbf{F}^{-T}\right] = \mathbf{0},$$

where $\mathbf{F} = \operatorname{Grad}\mathbf{r}$ is the *deformation gradient*, we transform Eqs. (3.30) and (3.31) to the *Lagrangian motion equations*

$$\rho_0 \frac{d\mathbf{v}}{dt} = \operatorname{Div}\mathbf{T}_\kappa + \rho_0 \mathbf{f}, \quad \rho_0 \gamma \frac{d\boldsymbol{\omega}}{dt} = \operatorname{Div}\mathbf{M}_\kappa - \left(\mathbf{T}_\kappa \cdot \mathbf{F}^T\right)_\times + \rho_0 \mathbf{m} \qquad (3.39)$$

with boundary conditions

$$\mathbf{T}_\kappa \cdot \mathbf{N} = \mathbf{t}_0, \quad \mathbf{M}_\kappa \cdot \mathbf{N} = \boldsymbol{\mu}_0 \quad \text{on } \Sigma_f, \qquad \mathbf{r} = \mathbf{r}_*, \quad \mathbf{H} = \mathbf{H}_* \quad \text{on} \quad \Sigma_u, \qquad (3.40)$$

where $\rho_0 = \rho J$ is the mass density in the reference configuration, $\rho_0 \gamma = jJ$, $J = \det \mathbf{F}$,

$$\mathbf{T}_\kappa = J\,\mathbf{T} \cdot \mathbf{F}^{-T}, \quad \mathbf{M}_\kappa = J\,\mathbf{M} \cdot \mathbf{F}^{-T}$$

are the *first Piola-Kirchhoff stress tensor and the couple stress tensor,* respectively, Div is the divergence operator in the reference configuration, $\mathbf{t}_0 = \mathbf{t}_* \frac{\mathrm{d}\sigma}{\mathrm{d}\Sigma}$ and $\boldsymbol{\mu}_0 = \boldsymbol{\mu}_* \frac{\mathrm{d}\sigma}{\mathrm{d}\Sigma}$ are given external surface loads acting on the surface in the reference configuration, and $\mathbf{N}$ is the unit normal to the boundary surface in the reference configuration.

The equilibrium of the micropolar continuum is described by the Eulerian equilibrium equations

$$\operatorname{div} \mathbf{T} + \rho \mathbf{f} = \mathbf{0}, \quad \operatorname{div} \mathbf{M} - \mathbf{T}_\times + \rho \mathbf{m} = \mathbf{0}, \qquad (3.41)$$

or by the Lagrangian equilibrium equations

$$\operatorname{Div} \mathbf{T}_\kappa + \rho_0 \mathbf{f} = \mathbf{0}, \quad \operatorname{Div} \mathbf{M}_\kappa - \left(\mathbf{T}_\kappa \cdot \mathbf{F}^T\right)_\times + \rho_0 \mathbf{m} = \mathbf{0}, \qquad (3.42)$$

complemented by the boundary conditions (3.34) and (3.35) or (3.40), respectively.

In linear micropolar elasticity the existence of solution and some properties of solutions of boundary-value problems are established in [7–9]. Mathematical studies of micropolar hydrodynamics can be found in [10]. The study of qualitative questions of nonlinear micropolar continuum is far from completion.

References

1. C. Truesdell, W. Noll, The nonlinear field theories of mechanics. in *Handbuch der Physik*, vol. III/3, ed. by S. Flügge (Springer, Berlin, 1965), pp. 1–602
2. C. Truesdell, *A First Course in Rational Continuum Mechanics* (Academic Press, New York, 1977)
3. C. Truesdell, *Rational Thermodynamics*, 2nd edn. (Springer, New York, 1984)
4. L.P. Lebedev, M.J. Cloud, V.A. Eremeyev, *Tensor Analysis with Applications in Mechanics* (World Scientific, New Jersey, 2010)
5. R.A. Horn, C.R. Johnson, *Matrix Analysis* (Cambridge University Press, Cambridge, 1985)
6. N.P. Migoun, P.P. Prokhorenko, *Hydrodynamics and Heattransfer in Gradient Flows of Microstructured Fluids (in Russian)* (Nauka i Technika, Minsk, 1984)
7. D. Iesan, Existence theorems in the theory of micropolar elasticity. Int. J. Eng. Sci. **8**, 777–791 (1970)
8. D. Iesan, Existence theorems in micropolar elastostatics. Int. J. Eng. Sci. **9**, 59–78 (1971)
9. W. Nowacki, *Theory of Asymmetric Elasticity* (Pergamon-Press, Oxford, 1986)
10. G. Lukaszewicz, *Micropolar Fluids: Theory and Applications* (Birkhäuser, Boston, 1999)

Chapter 4
Constitutive Equations

For an arbitrary part of the body, Eqs. (3.30) and (3.31) express the balance equations for the moment and the moment of momentum. These six scalar equations contain 18 unknown quantities that are the components of tensors $\mathbf{T}$ and $\mathbf{M}$. The dependence of $\mathbf{T}$ and $\mathbf{M}$ on medium deformations is determined by the *constitutive equations* or *constitutive relations* that depend on the material properties. They are determined experimentally. The constitutive equations must obey some principles that restrict their form, see [1].

In the framework of micropolar continuum mechanics we introduce the constitutive equations for $\mathbf{T}$ and $\mathbf{M}$. Besides, we can introduce another type of constitutive equations that determines the moment and the moment of momentum of more general form than those given by definitions (3.1) and (3.2). These relations are called *kinetic constitutive equations*. Note that in the general nonlinear shell theory, the dynamic equations coincide with two-dimensional dynamic equations for the micropolar continuum with more complex relations for the inertia tensors.

4.1 General Principles Restricting the Constitutive Equations

In this section we formulate the principles for constitutive equations for micropolar continuum. We restrict them to the case of the pure mechanical theory. Here we will omit the influence of temperature and do not introduce internal energy, entropy, etc. The restrictions stipulated on the form of constitutive equations by the principles must be valid for any medium, cf. [1].

Principle of Determinism. *At each body point, both the stress and the couple stress tensors are uniquely determined by the pre-history of the body motion.*

Principle of Local Action. *At each body point, both the stress and the couple stress tensors are uniquely determined by the motion of any neighborhood of the point that can be so small as is desired.*

V. A. Eremeyev et al., *Foundations of Micropolar Mechanics*,
SpringerBriefs in Continuum Mechanics,
DOI: 10.1007/978-3-642-28353-6_4, © The Author(s) 2013

Let us note that the Principle of Determinism states that the constitutive equations cannot forecast the future and so it is absolute whereas the Principle of Local Action is not: it can be not valid for some materials. Non-local continuum models are presented for example in [2].

Principle of Material Frame-Indifference. *Being determined by the constitutive equations, the stress and couple stress tensors must be material frame-indifferent quantities.*

To introduce the notion of frame indifference we remind the definition of equivalent motions. In classical mechanics, two motions $\mathbf{r}$ and $\mathbf{r}^*$ are called *equivalent* if they relate as follows

$$\mathbf{r}^* = \mathbf{a}(t) + \mathbf{O}(t) \cdot (\mathbf{r} - \mathbf{r}_0), \tag{4.1}$$

where $\mathbf{O}(t)$ is an arbitrary orthogonal tensor, $\mathbf{a}(t)$ is an arbitrary vector function and the constant vector $\mathbf{r}_0$ represents a fixed point position (a pole). We can treat equivalent motions as the one body motion considered in different reference frames.

In micropolar mechanics, body deformations are described by the position vector $\mathbf{r}$ and the trihedron $\mathbf{d}_k$ or the orthogonal tensor $\mathbf{H}$. After [3, 4], we *assume* that in the equivalent motion the directors $\mathbf{d}_k$ rotate similarly to $\mathbf{r}$:

$$\mathbf{d}_k^* = \mathbf{O}(t) \cdot \mathbf{d}_k.$$

It follows that in the equivalent motion the microrotation tensors relate by the equation:

$$\mathbf{H}^* = \mathbf{O}(t) \cdot \mathbf{H}. \tag{4.2}$$

We underline that two deformations of the micropolar medium are equivalent if the position vectors and microrotation tensors relate through Eqs. (4.1) and (4.2).

After [1] we recall that the scalar x, the vector $\mathbf{x}$ and the tensor $\mathbf{X}$ are called indifferent or objective if for two equivalent motions there hold

$$\begin{aligned} x^* &= x, \\ \mathbf{x}^* &= \mathbf{O} \cdot \mathbf{x}, \\ \mathbf{X}^* &= \mathbf{O} \cdot \mathbf{X} \cdot \mathbf{O}^T, \end{aligned} \tag{4.3}$$

where superscript "$*$" denotes the quantities in the equivalent motion. It is obvious that the directors $\mathbf{d}_k$ are objective vectors.

As an example let us consider the *mirror reflection* that is when $\mathbf{O} = -\mathbf{I}$. From (4.3) it follows the relations

$$x^* = x, \quad \mathbf{x}^* = -\mathbf{x}, \quad \mathbf{X}^* = \mathbf{X}.$$

The principle of material frame-indifference or the principle of objectivity was originally proposed for classical continuum mechanics by Noll [5], see [1]. There has been an extensive discussion in the literature about the proper understanding of this principle, because its different formulations seem to reflect different physical contents, see for example recent papers by Murdoch [6–8], Muschik and Restuccia [9], Rivlin [10–12], Bertram and Svendsen [13–15], and the book [16]. In particular, in [15] it is stated that the principle of material frame-indifference contains three principles

- the principle of invariance under Euclidean transforms,
- the principle of invariance under superimposed rigid-body motions,
- the principle of form-invariance of the constitutive equations under change of the observer.

If any two of them are satisfied the third one becomes satisfied as well.

For micropolar medium, the material frame-indifference principle needs to be specified. Unlike classical mechanics that uses only polar vectors and tensors, micropolar mechanics employs axial vectors and tensors that are not "real" frame-indifferent quantities. Under mirror reflection, the change of sign of axial quantities differs from the one for polar quantities. We discuss the difference in Appendix A in more details. If in the definition of the equivalent motion we leave only proper orthogonal vectors $\mathbf{O}$ there is no difference between axial and polar vectors. However non-proper orthogonal tensors correspond to the rotation of coordinates and to the coordinate orientation change. Such transformations look admissible but axial quantities have quite certain physical meaning as well.

The definition for frame-indifferent axial quantities transforms as follows. The axial scalar (pseudoscalar) ω, the axial vector $\boldsymbol{\omega}$ and the axial tensor $\boldsymbol{\Omega}$ are called indifferent or objective if for two equivalent motions there hold

$$\begin{aligned} \omega^* &= (\det \mathbf{O})\omega, \\ \boldsymbol{\omega}^* &= (\det \mathbf{O})\mathbf{O} \cdot \boldsymbol{\omega}, \\ \boldsymbol{\Omega}^* &= (\det \mathbf{O})\mathbf{O} \cdot \boldsymbol{\Omega} \cdot \mathbf{O}^T. \end{aligned} \tag{4.4}$$

In particular, for mirror reflection $\mathbf{O} = -\mathbf{I}$ and

$$\omega^* = -\omega, \quad \boldsymbol{\omega}^* = \boldsymbol{\omega}, \quad \boldsymbol{\Omega}^* = -\boldsymbol{\Omega}.$$

Among the introduced quantities the total moment, the couple stress vector $\boldsymbol{\mu}$, and the couple stress tensor $\mathbf{M}$ are axial ones.

At first glance the transformation rule (4.4) contradicts to a naive understanding of independence of an objective quantity on the choice of coordinates. However, it is not correct. The dependence on mirror reflections is a consequence of the agreement on the choice of right-hand or left-hand rules when we define such quantities as the cross product or the couple. It is similar to the agreement of what of the Earth poles is northern or what of the rotation directions we consider to be positive. Described by indifferent axial quantities, physical processes do not depend on the choice of coordinates, in particular they do not depend on the change of the frame orientation.

An interesting discussion of the physical nature of axial vectors and mirror reflection can be found in [17].

4.2 Natural Lagrangian Strain Measures of Cosserat Continuum

Following Pietraszkiewicz and Eremeyev [18, 19] we discuss here the definitions of the Lagrangian stretch and wryness tensors in the non-linear Cosserat continuum. The stretch and wryness tensors of the non-linear Cosserat continuum were originally defined by Cosserats [20] through components of some fields in the common Cartesian frame. We present three different methods of introducing the strain measures into the non-linear Cosserat continuum:

(a) by a direct geometrical approach,
(b) by defining the strain measures as the fields work-conjugate to the respective internal stress and couple-stress tensor fields,
(c) by applying the principle of material frame-indifference to the strain energy density of the polar-elastic body.

Each of the three ways allows to associate different geometrical and/or physical interpretations to the corresponding strain measures. The strain measures of [18] called the *natural* ones are of the *relative type*. They have to vanish in the reference configuration.

4.2.1 Strain Measures by Geometrical Approach

Within the geometrical approach we define the strain measures by analyzing the difference of the fields describing position and orientation differentials of the material particles of the micropolar continuum in the 3D physical space.

Since $\mathbf{D}_k$ are the unit vectors, the vector $\mathrm{d}\mathbf{D}_k$ can be represented by the axial vector $\mathbf{b}$ depending linearly on $\mathrm{d}\mathbf{R}$, so that

$$\mathrm{d}\mathbf{D}_k = \mathbf{b} \times \mathbf{D}_k, \quad \mathbf{b} = \mathbf{B} \cdot \mathrm{d}\mathbf{R}, \quad \mathbf{b} = \frac{1}{2}\mathbf{D}_k \times \mathrm{d}\mathbf{D}_k, \quad \mathbf{B} = \frac{1}{2}\mathbf{D}_k \times \operatorname{Grad}\mathbf{D}_k. \quad (4.5)$$

In (4.5), $\mathbf{B}$ is the *microstructure curvature tensor* in the reference configuration of the micropolar continuum. Two tensors $\mathbf{I}, \mathbf{B}$ are the basic measures of local geometry of the reference configuration κ.

Again, $\mathrm{d}\mathbf{d}_k$ can be represented by the axial vector $\mathbf{c}$ depending linearly on $\mathrm{d}\mathbf{r}$ by the relations

$$\mathrm{d}\mathbf{d}_k = \mathbf{c} \times \mathbf{d}_k, \quad \mathbf{c} = \mathbf{C} \cdot \mathrm{d}\mathbf{r}, \quad \mathbf{c} = \frac{1}{2}\mathbf{d}_k \times \mathrm{d}\mathbf{d}_k, \quad \mathbf{C} = \frac{1}{2}\mathbf{d}_k \times \operatorname{grad}\mathbf{d}_k. \quad (4.6)$$

where $\mathbf{C}$ is the *microstructure curvature tensor* in the actual configuration of the micropolar continuum. Two tensors $\mathbf{I}$ and $\mathbf{C}$ are the basic measures of local geometry of the actual configuration χ.

In the actual configuration χ the differential of $\mathbf{r}$ is

$$\mathrm{d}\mathbf{r} = (\mathrm{grad}\,\mathbf{r}) \cdot \mathrm{d}\mathbf{r} = (\mathrm{Grad}\,\mathbf{r}) \cdot \mathrm{d}\mathbf{R} = \mathbf{F} \cdot \mathrm{d}\mathbf{R}, \tag{4.7}$$

where grad denotes the gradient operator in χ, $\mathrm{grad}\,\mathbf{r} = \mathbf{I}$, and $\mathbf{F} = \mathrm{Grad}\,\mathbf{r}$ is the classical *deformation gradient* tensor.

Since $\mathbf{H}^T \cdot \mathbf{H}_{,i} = -(\mathbf{H}^T \cdot \mathbf{H}_{,i})^T$ is skew-symmetric it can be expressed through the axial vector $\mathbf{k}_i$, see (A.5) and (A.6),

$$\mathbf{H}^T \cdot \mathbf{H}_{,i} = \mathbf{k}_i \times \mathbf{I} = \mathbf{I} \times \mathbf{k}_i, \quad \mathbf{k}_i = -\frac{1}{2}(\mathbf{H}^T \cdot \mathbf{H}_{,i})_\times, \tag{4.8}$$

where $(\ldots)_{,i}$, $i = 1, 2, 3$, denotes the derivative with respect to the Cartesian coordinate x_i in the reference configuration. This allows one to introduce the second-order tensor

$$\mathbf{K} = \mathbf{k}_j \otimes \mathbf{i}_j = -\frac{1}{2}(\mathbf{H}^T \cdot \mathbf{H}_{,j})_\times \otimes \mathbf{i}_j = -\frac{1}{2}\mathcal{E} : \left(\mathbf{H}^T \cdot \mathrm{Grad}\,\mathbf{H}\right), \qquad \mathbf{H}^T \cdot \mathrm{Grad}\,\mathbf{H} = \mathbf{I} \times \mathbf{K}, \tag{4.9}$$

where the double dot product : of two third-order tensors $\mathbb{A}$, $\mathbb{B}$ represented in the base $\mathbf{i}_k$ is defined as $\mathbb{A} : \mathbb{B} = A_{amn} B_{mnb} \mathbf{i}_a \otimes \mathbf{i}_b$. The tensor $\mathbf{K}$ characterizes uniquely the third-order tensor $\mathbf{H}^T \cdot \mathrm{Grad}\,\mathbf{H}$ skew-symmetric with regard to first two tensor places. The tensor $\mathbf{K}$ is frequently called the *wryness tensor* in the literature, cf. [3].

Using the chain rule $\mathrm{grad}\,\mathbf{d}_k = (\mathrm{Grad}\,\mathbf{d}_k) \cdot \mathbf{F}^{-1}$ with (4.8) and (4.9), the tensor $\mathbf{C}$ can now be represented by

$$\mathbf{C} = \mathbf{H} \cdot (\mathbf{K} + \mathbf{B}) \cdot \mathbf{F}^{-1}. \tag{4.10}$$

The relative changes of lengths and orientations of the micropolar continuum during deformation are governed by differences of differentials (4.5)–(4.7) brought to the comparable orientation by the tensor $\mathbf{H}$,

$$\mathrm{d}\mathbf{r} - \mathbf{H} \cdot \mathrm{d}\mathbf{R} = \mathbf{X} \cdot \mathrm{d}\mathbf{R} = \mathbf{e} \cdot \mathrm{d}\mathbf{r}, \quad \mathbf{C} \cdot \mathrm{d}\mathbf{r} - \mathbf{H} \cdot \mathbf{B} \cdot \mathrm{d}\mathbf{R} = \boldsymbol{\Phi} \cdot \mathrm{d}\mathbf{R} = \mathbf{k} \cdot \mathrm{d}\mathbf{r}, \tag{4.11}$$

$$\mathbf{X} = \mathbf{F} - \mathbf{H}, \quad \mathbf{e} = \mathbf{I} - \mathbf{H} \cdot \mathbf{F}^{-1} = \mathbf{X} \cdot \mathbf{F}^{-1}, \tag{4.12}$$

$$\boldsymbol{\Phi} = \mathbf{C} \cdot \mathbf{F} - \mathbf{H} \cdot \mathbf{B}, \quad \mathbf{k} = \mathbf{C} - \mathbf{H} \cdot \mathbf{B} \cdot \mathbf{F}^{-1} = \boldsymbol{\Phi} \cdot \mathbf{F}^{-1}. \tag{4.13}$$

Scalar products of each of (4.11) by itself leads to the quadratic forms

$$\mathrm{d}\mathbf{R}\cdot\mathbf{X}^T\cdot\mathbf{X}\cdot\mathrm{d}\mathbf{R} = \mathrm{d}\mathbf{r}\cdot\mathbf{e}^T\cdot\mathbf{e}\cdot\mathrm{d}\mathbf{r}, \quad \mathrm{d}\mathbf{R}\cdot\boldsymbol{\Phi}^T\cdot\boldsymbol{\Phi}\cdot\mathrm{d}\mathbf{R} = \mathrm{d}\mathbf{r}\cdot\mathbf{k}^T\cdot\mathbf{k}\cdot\mathrm{d}\mathbf{r}. \quad (4.14)$$

However, the relative changes of lengths and orientations can also be calculated by the alternative back-rotated expressions

$$\mathbf{H}^T\cdot\mathrm{d}\mathbf{r}-\mathrm{d}\mathbf{R} = \mathbf{E}\cdot\mathrm{d}\mathbf{R} = \mathbf{x}\cdot\mathrm{d}\mathbf{r}, \quad \mathbf{H}^T\cdot\mathbf{C}\cdot\mathrm{d}\mathbf{r}-\mathbf{B}\cdot\mathrm{d}\mathbf{R} = \mathbf{K}\cdot\mathrm{d}\mathbf{R} = \boldsymbol{\varphi}\cdot\mathrm{d}\mathbf{r}, \quad (4.15)$$

$$\mathbf{E} = \mathbf{H}^T\cdot\mathbf{F}-\mathbf{I} = \mathbf{H}^T\cdot\mathbf{X}, \quad (4.16)$$

$$\mathbf{x} = \mathbf{H}^T-\mathbf{F}^{-1} = \mathbf{E}\cdot\mathbf{F}^{-1} = \mathbf{H}^T\cdot\mathbf{e} = \mathbf{H}^T\cdot\mathbf{X}\cdot\mathbf{F}^{-1}, \quad (4.17)$$

$$\mathbf{K} = \mathbf{H}^T\cdot\mathbf{C}\cdot\mathbf{F}-\mathbf{B} = \mathbf{H}^T\cdot\boldsymbol{\Phi}, \quad (4.18)$$

$$\boldsymbol{\varphi} = \mathbf{H}^T\cdot\mathbf{C}-\mathbf{B}\cdot\mathbf{F}^{-1} = \mathbf{K}\cdot\mathbf{F}^{-1} = \mathbf{H}^T\cdot\mathbf{k} = \mathbf{H}^T\cdot\boldsymbol{\Phi}\cdot\mathbf{F}^{-1}. \quad (4.19)$$

From (4.9), $(4.12)_2$, $(4.13)_2$, (4.16), and (4.18) we obtain the following relations:

$$\mathbf{e} = \mathbf{H}\cdot\mathbf{E}\cdot\mathbf{F}^{-1}, \quad \mathbf{k} = \mathbf{H}\cdot\mathbf{K}\cdot\mathbf{F}^{-1} = -\frac{1}{2}\mathbf{H}\cdot\mathcal{E} : (\mathbf{H}^T\cdot\operatorname{grad}\mathbf{H}). \quad (4.20)$$

Scalar products of each of (4.15) by itself give the alternative quadratic forms

$$\mathrm{d}\mathbf{R}\cdot\mathbf{E}^T\cdot\mathbf{E}\cdot\mathrm{d}\mathbf{R} = \mathrm{d}\mathbf{r}\cdot\mathbf{x}^T\cdot\mathbf{x}\cdot\mathrm{d}\mathbf{r}, \quad \mathrm{d}\mathbf{R}\cdot\mathbf{K}^T\cdot\mathbf{K}\cdot\mathrm{d}\mathbf{R} = \mathrm{d}\mathbf{r}\cdot\boldsymbol{\varphi}^T\cdot\boldsymbol{\varphi}\cdot\mathrm{d}\mathbf{r}. \quad (4.21)$$

From (4.14) and (4.21) it follows that each of the tensors $\mathbf{X}$, $\mathbf{E}$, or $\mathbf{x}$, $\mathbf{e}$ and $\boldsymbol{\Phi}$, $\mathbf{K}$ or $\boldsymbol{\varphi}$, $\mathbf{k}$ is the corresponding measure of deformation, stretch or orientation change of the non-linear micropolar continuum in the Lagrangian or Eulerian description, respectively.

The quadratic forms (4.14) and (4.21) do not change if $\mathbf{X}$, $\mathbf{E}$, $\boldsymbol{\Phi}$, $\mathbf{K}$ and their counterparts are replaced by $\mathbf{R}\cdot\mathbf{X}$, $\mathbf{R}\cdot\mathbf{E}$, $\mathbf{R}\cdot\boldsymbol{\Phi}$, $\mathbf{R}\cdot\mathbf{K}$, etc., respectively, where $\mathbf{R}$ is a orthogonal tensor. Hence, any so transformed tensor can also be regarded as the possible strain measure of the non-linear micropolar continuum. In particular when such a transformation with $\mathbf{R} = \mathbf{H}^T$ is applied to the measures $\mathbf{X}$, $\mathbf{e}$, $\boldsymbol{\Phi}$, $\mathbf{k}$ entering the quadratic form (4.14) the measures become $\mathbf{E}$, $\mathbf{x}$, $\mathbf{K}$, $\boldsymbol{\varphi}$, i.e. those entering the quadratic form (4.21).

It follows from (4.12), (4.13), and (4.17) that $\mathbf{X}$, $\boldsymbol{\Phi}$ (and $\mathbf{x}$, $\boldsymbol{\varphi}$) are two-point tensors with the left leg associated with the actual configuration and the right leg with the reference one (and reverse for $\mathbf{x}$, $\boldsymbol{\varphi}$). Such measures may also be called the deformation measures. The tensors $\mathbf{e}$, $\mathbf{k}$ are the *relative Eulerian strain measures*, while the tensors $\mathbf{E}$, $\mathbf{K}$ are the *relative Lagrangian strain measures*, or the *Lagrangian stretch and wryness tensors*, respectively. The latter are given by the formulae

$$\mathbf{E} = \mathbf{H}^T \cdot \mathbf{F} - \mathbf{I}, \quad \mathbf{K} = \mathbf{H}^T \cdot \mathbf{C} \cdot \mathbf{F} - \mathbf{B} = -\frac{1}{2}\mathcal{E} : (\mathbf{H}^T \cdot \operatorname{Grad} \mathbf{H}). \qquad (4.22)$$

Let us note some interesting features of the relative Lagrangian strain measures:

1. They are given in the common coordinate-free notation; their various component representations can easily be generated, if necessary.
2. Definitions of the measures are valid for finite translations and rotations as well as for unrestricted stretches and changes of the microstructure orientation of the Cosserat body.
3. The measures are expressed in terms of the rotation tensor $\mathbf{H}$; for any specific parametrization of the rotation group $SO(3)$ by various finite rotation vectors, Euler angles, quaternions, etc. appropriate expressions for the measures can easily be found, if necessary.
4. The measures vanish in the rigid-body deformation $\mathbf{r} = \mathbf{O} \cdot \mathbf{R} + \mathbf{a}$, $\mathbf{d}_k = \mathbf{O} \cdot \mathbf{D}_k$ with a constant vector $\mathbf{a}$ and a constant proper orthogonal tensor $\mathbf{O}$ defined for the whole body.
5. In the absence of deformation from the reference configuration, that is when $\mathbf{F} = \mathbf{H} = \mathbf{I}$, the measures identically vanish.
6. The measures are not symmetric, in general: $\mathbf{E}^T \neq \mathbf{E}$, $\mathbf{K}^T \neq \mathbf{K}$.

In this purely geometrical approach there is no need for discussion whether these measures might be defined as transposed ones or with opposite signs. Elements of geometrical approach in Cartesian components were used already by Cosserats [20] and more recently by Merlini [21] who took explicitly into account the microstructure curvature tensors describing spatial changes of orientations of the material particles in the reference and actual configurations. These tensors are independently introduced also in [22, 23] within the theory of viscoelastic micropolar fluids, and in [24] within the general theory of shells. The microstructure curvature tensors are extensively used in discussions on the local material symmetry group of elastic shells and polar-elastic media in [25, 26].

4.2.2 Principle of Virtual Work and Work-Conjugate Strain Measures

Already Reissner [27, 28] noted that the internal structure of two local equilibrium equations of the Cosserat elastic body requires two specific strain measures expressed in terms of independent translation and rotation vectors as the only field variables. This idea is presented here in the general case of the non-linear Cosserat continuum using the coordinate-free approach.

If the Lagrangian equilibrium conditions of the Cosserat continuum (3.42) are multiplied by two arbitrary smooth vector fields $\mathbf{v}$, $\boldsymbol{\omega}$, then we generate the integral identity

$$\iiint_V \{(\mathrm{Div}\,\mathbf{T}_\kappa + \rho_0\mathbf{f})\cdot\mathbf{v} + [\mathrm{Div}\,\mathbf{M}_\kappa - (\mathbf{T}_\kappa\cdot\mathbf{F}^T)_\times + \rho_0\mathbf{m}]\cdot\boldsymbol{\omega}\}\,\mathrm{d}V \tag{4.23}$$
$$-\iint_{\Sigma_f}\{(\mathbf{T}_\kappa\cdot\mathbf{N}-\mathbf{t}_0)\cdot\mathbf{v} + (\mathbf{M}_\kappa\cdot\mathbf{N}-\boldsymbol{\mu}_0)\cdot\boldsymbol{\omega}\}\quad \mathrm{d}\Sigma = 0.$$

The vector field $\mathbf{v}$ can be interpreted as the kinematically admissible virtual translation $\mathbf{v}\equiv\delta\mathbf{r}$ and the vector field $\boldsymbol{\omega}$ as the kinematically admissible virtual rotation $\boldsymbol{\omega}\equiv -\frac{1}{2}\left(\mathbf{H}^T\cdot\delta\mathbf{H}\right)_\times$ such that $\mathbf{v}=\boldsymbol{\omega}=\mathbf{0}$ on Σ_u, where δ is the symbol of virtual change (variation). Then using the divergence theorem (A.35)$_2$ the identity (4.23) can be transformed into the *principle of virtual work*

$$\iiint_V \left[\mathbf{T}_\kappa\cdot(\mathrm{Grad}\,\mathbf{v}-\boldsymbol{\omega}\times\mathbf{F}) + \mathbf{M}_\kappa\cdot\mathrm{Grad}\,\boldsymbol{\omega}\right]\,\mathrm{d}V \tag{4.24}$$
$$=\iiint_V \rho_0\,(\mathbf{f}\cdot\mathbf{v}+\mathbf{m}\cdot\boldsymbol{\omega})\;\mathrm{d}V + \iint_{\Sigma_f}(\mathbf{t}_0\cdot\mathbf{v}+\boldsymbol{\mu}_0\cdot\boldsymbol{\omega})\;\mathrm{d}\Sigma.$$

But we can show that $\delta\mathbf{E}=\mathbf{H}^T\cdot(\mathrm{Grad}\,\mathbf{v}-\boldsymbol{\omega}\times\mathbf{F})$ and $\delta\mathbf{K}=\mathbf{H}^T\cdot\mathrm{Grad}\,\boldsymbol{\omega}$, and the *internal virtual work density* under the first volume integral of (4.24) can now be given by the expressions

$$\sigma = \mathbf{T}_\kappa\cdot(\mathbf{H}\cdot\delta\mathrm{E}) + \mathbf{M}_\kappa\cdot(\mathbf{H}\cdot\delta\mathbf{K}) = \mathbf{P}_1\cdot\delta\mathbf{E} + \mathbf{P}_2\cdot\delta\mathbf{K}, \tag{4.25}$$

where $\mathbf{P}_1=\mathbf{H}^T\cdot\mathbf{T}_\kappa$, $\mathbf{P}_2=\mathbf{H}^T\cdot\mathbf{M}_\kappa$ are the non-symmetric stress and couple-stress tensors whose natural components are referred entirely to the reference configuration. We call $\mathbf{P}_1$ and $\mathbf{P}_2$ the *second Piola–Kirchhoff stress and couple stress tensors*, respectively. The stress measures $\mathbf{P}_1$, $\mathbf{P}_2$ require the relative Lagrangian strain measures $\mathbf{E}$, $\mathbf{K}$ as their work-conjugate counterparts.

4.2.3 Invariance of the Polar-Elastic Strain Energy Density

In the polar-elastic body the constitutive relations are defined through the strain energy density W_κ per unit volume of the reference configuration κ. In general, the density W_κ can be assumed in the following form:

$$W_\kappa = W_\kappa(\mathbf{r},\mathbf{F},\mathbf{H},\mathrm{Grad}\,\mathbf{H};\mathbf{R},\mathbf{B}). \tag{4.26}$$

But W_κ in (4.26) should satisfy the principle of material frame-indifference or invariance under the superposed rigid-body deformations. Applying (4.1) and (4.2) we obtain that the principle of material frame-indifference requires the values of W_κ to be the same for both deformations $\mathbf{r}$, $\mathbf{H}$ and $\mathbf{r}^*$, $\mathbf{H}^*$,

$$W_\kappa(\mathbf{r},\ \mathbf{F},\ \mathbf{H},\ \mathrm{Grad}\,\mathbf{H};\ \mathbf{R},\mathbf{B}) = W_\kappa(\mathbf{O}\cdot\mathbf{r}+\mathbf{a},\mathbf{O}\cdot\mathbf{F},\mathbf{O}\cdot\mathbf{H},\mathbf{O}\cdot\mathrm{Grad}\,\mathbf{H};\ \mathbf{R},\mathbf{B}). \quad (4.27)$$

Since $\mathbf{a}$ is arbitrary, in order to assure invariance of W_κ in (4.27) the density should not depend on $\mathbf{r}$. Then, if $\mathbf{O} \equiv \mathbf{H}^T$, the function W_κ can be reduced to

$$W_\kappa = W_\kappa(\mathbf{H}^T\cdot\mathbf{F},\mathbf{I},\mathbf{H}^T\cdot\mathrm{Grad}\,\mathbf{H};\ \mathbf{R},\mathbf{B}), \quad (4.28)$$

which by $(4.9)_2$ and $(4.22)_1$ becomes equivalent to

$$W_\kappa = W_\kappa\,(\mathbf{E}+\mathbf{I},\ \mathbf{I},\mathbf{I}\times\mathbf{K};\ \mathbf{R},\mathbf{B}) = \overline{W}_\kappa\,(\mathbf{E},\mathbf{K};\ \mathbf{R},\mathbf{B}). \quad (4.29)$$

This again confirms that the relative Lagrangian strain measures $\mathbf{E}$, $\mathbf{K}$ are required to be the independent fields in the polar-elastic strain energy density in order it to be invariant under the superposed rigid-body deformation. This way of introducing the Lagrangian strain measures is most common in the literature and various such procedures are used, for example, in [3, 22, 29–33].

Since for the micropolar elastic continuum $\sigma = \delta W_\kappa$ from (4.25) it follows the constitutive equations for the stress measures

$$\mathbf{P}_1 = \frac{\partial W_\kappa}{\partial \mathbf{E}},\quad \mathbf{P}_2 = \frac{\partial W_\kappa}{\partial \mathbf{K}},\quad \mathbf{T}_\kappa = \mathbf{H}\cdot\frac{\partial W_\kappa}{\partial \mathbf{E}},\quad \mathbf{M}_\kappa = \mathbf{H}\cdot\frac{\partial W_\kappa}{\partial \mathbf{K}}. \quad (4.30)$$

The geometrical approach, the structure of equilibrium conditions and the invariance of the polar-elastic strain energy density all require the tensors $\mathbf{E}$, $\mathbf{K}$ as the most appropriate Lagrangian strain measures for the non-linear Cosserat continuum. We call the measures the *natural* stretch and wryness tensors, respectively.

In [18] is shown that the stretch and wryness tensors introduced in many papers do not agree with each other and with the Lagrangian strain measures defined in (4.22). Most definitions differ only by transpose of the measures or by opposite signs. Some measures do not vanish in the absence of deformation. Such differences are not essential for the theory, although one should be aware of them.

4.2.4 Vectorial Parameterizations of Strain Measures

While three components of $\mathbf{u}$ in (4.22) are all independent, nine components of $\mathbf{H}$ in (4.22) are subjected to six constraints following from the orthogonality conditions $\mathbf{H}^{-1} = \mathbf{H}^T$, $\det\mathbf{H} = +1$, so that only three rotational parameters of $\mathbf{H}$ are independent. In many applications it is more convenient to use the strain measures expressed in terms of six disconfiguration parameters all of which are independent.

Many techniques how to parameterize the rotation group $SO(3)$ are developed, which can roughly be classified as vectorial and non-vectorial ones. Various finite rotation vectors as well as the Cayley–Gibbs and exponential map parameters are examples of the vectorial parameterization, for they all have three independent scalar parameters as Cartesian components of a generalized vector in the 3D vector space. The non-vectorial parameterizations are expressed either in terms of three scalar parameters that cannot be treated as vector components, such as Euler-type angles for example, or through more scalar parameters subject to additional constraints, such as unit quaternions, Cayley–Klein parameters, or direction cosines. Each of these expressions may appear to be more convenient than others when solving specific problems of the non-linear Cosserat continuum.

The microrotation tensor $\mathbf{H}$ can be expressed by the Gibbs formula (A.7). In the vectorial parameterization of $\mathbf{H}$ one introduces a scalar function $p(\phi)$ generating three components of the *finite rotation vector* $\mathbf{p}$ defined as $\mathbf{p} = p(\phi)\mathbf{e}$, see for example [34]. The generating function $p(\phi)$ has to be an odd function of ϕ with the limit behavior

$$\lim_{\phi\to 0} \frac{p(\phi)}{\phi} = \kappa,$$

where κ is a positive real normalization factor (usually 1 or $\frac{1}{2}$), and $p(0) = 0$. Then the tensor $\mathbf{H}$ can be represented as

$$\mathbf{H} = \cos\phi\mathbf{I} + \frac{1-\cos\phi}{p^2}\mathbf{p}\otimes\mathbf{p} - \frac{\sin\phi}{p}\mathbf{p}\times\mathbf{I}. \tag{4.31}$$

Taking the gradient of (4.31) and substituting it into (4.22), after appropriate transformations the natural Lagrangian stretch $\mathbf{E}$ and wryness $\mathbf{K}$ tensors can be represented in terms of the finite rotation vector $\mathbf{p}$ and the translation vector $\mathbf{u}$ by the relations

$$\mathbf{E} = \left(\cos\phi\mathbf{I} + \frac{1-\cos\phi}{p^2}\mathbf{p}\otimes\mathbf{p} + \frac{\sin\phi}{p}\mathbf{p}\times\mathbf{I}\right)(\mathbf{I} + \operatorname{Grad}\mathbf{u}) - \mathbf{I}, \tag{4.32}$$

$$\mathbf{K} = \left[\frac{-\sin\phi}{p}\mathbf{I} + \frac{1}{p^2}\left(\frac{1}{p'} + \frac{\sin\phi}{p}\right)\mathbf{p}\otimes\mathbf{p} - \frac{1-\cos\phi}{p^2}\mathbf{p}\times\mathbf{I}\right]\operatorname{Grad}\mathbf{p}. \tag{4.33}$$

Among definitions of $\mathbf{p}$ used in the literature let us mention the finite rotation vectors defined as

$$\boldsymbol{\theta} = 2\tan\frac{\phi}{2}\,\mathbf{e},\quad \boldsymbol{\varphi} = \phi\,\mathbf{e},\quad \bar{\boldsymbol{\omega}} = \sin\phi\,\mathbf{e},\quad \boldsymbol{\rho} = \tan\frac{\phi}{2}\,\mathbf{e}, \tag{4.34}$$

$$\boldsymbol{\sigma} = 2\sin\frac{\phi}{2}\,\mathbf{e}, \quad \boldsymbol{\mu} = 4\tan\frac{\phi}{4}\,\mathbf{e}, \quad \boldsymbol{\beta} = 4\sin\frac{\phi}{4}\,\mathbf{e}, \tag{4.35}$$

where the generating functions are

$$\theta = 2\tan\frac{\phi}{2}, \quad \phi, \varpi = \sin\phi, \quad \rho = \tan\frac{\phi}{2},$$
$$\sigma = 2\sin\frac{\phi}{2}, \quad \mu = 4\tan\frac{\phi}{4}, \text{ and } \beta = 4\sin\frac{\phi}{4},$$

respectively. The explicit formulae for E and K expressed in terms of the corresponding finite rotation vectors (4.34) and (4.35) are summarized in [19].

4.3 Kinetic Constitutive Equations

Deriving expressions for momentum and moment of momentum of an arbitrary body part, that is (3.1) and (3.2), we have used the simplest relations for linear and angular velocities $\mathbf{v}$ and $\boldsymbol{\omega}$. However, even for a rigid body we see that the momentum and moment of momentum can have a complicated form. Let us determine $\mathfrak{P}$ and $\mathfrak{M}$ as follows

$$\mathfrak{P}(\mathscr{P}) \triangleq \iiint\limits_{v_{\mathscr{P}}} \rho \mathbf{K}_1 \, dv, \quad \mathfrak{M}(\mathscr{P}) \triangleq \iiint\limits_{v_{\mathscr{P}}} \rho \left\{(\mathbf{r} - \mathbf{r}_0) \times \mathbf{K}_1 + \mathbf{K}_2\right\} dv, \tag{4.36}$$

where

$$\mathbf{K}_1 = \mathbf{K}_1(\mathbf{v}, \boldsymbol{\omega}; \mathbf{r}, \mathsf{H}, \operatorname{Grad}\mathbf{r}, \operatorname{Grad}\mathsf{H}), \quad \mathbf{K}_2 = \mathbf{K}_2(\mathbf{v}, \boldsymbol{\omega}; \mathbf{r}, \mathsf{H}, \operatorname{Grad}\mathbf{r}, \operatorname{Grad}\mathsf{H}). \tag{4.37}$$

These are the expressions for the density of momentum and moment of momentum of micropolar continuum. Such constitutive equations are called *kinetic*. Note that up to now we have dealt with the constitutive equation for frame-indifferent quantities like the strain energy density W. That allowed us to use the material frame-indifference principle to reduce the number of arguments in W and to get indifferent constitutive equations. Now $\mathbf{K}_1$ and $\mathbf{K}_2$ are not frame-indifferent quantities that is a consequence of the fact that $\mathbf{v}$ and $\boldsymbol{\omega}$ are not frame-indifferent. So we cannot use the material frame-indifference principle to simplify Eq. (4.37).

A rational way to simplify (4.37) is to use the analogy with the rigid body. Let us assume that $\mathbf{K}_1$ and $\mathbf{K}_2$ are linear functions of $\mathbf{v}$ and $\boldsymbol{\omega}$:

$$\mathbf{K}_1 = \mathsf{J}_1 \cdot \mathbf{v} + \mathsf{J}_2 \cdot \boldsymbol{\omega}, \quad \mathbf{K}_2 = \mathsf{J}_3 \cdot \mathbf{v} + \mathsf{J}_4 \cdot \boldsymbol{\omega},$$

where J_k, $k = 1, 2, 3, 4$, are *tensors of inertia*. In general, J_k are tensorial functions of arguments $\mathbf{r}$, H, $\operatorname{Grad}\mathbf{r}$, and $\operatorname{Grad}\mathsf{H}$.

For micropolar medium, it is natural to postulate the existence of the kinetic energy density $K = K(\mathbf{v}, \boldsymbol{\omega}; \mathbf{r}, \mathbf{H}, \operatorname{Grad}\mathbf{r}, \operatorname{Grad}\mathbf{H})$ that is the potential for $\mathbf{K}_1$ and $\mathbf{K}_2$. Now we get

$$K = \frac{1}{2}(\mathbf{v}\cdot\mathbf{J}_1\cdot\mathbf{v}+\mathbf{v}\cdot\mathbf{J}_2\cdot\boldsymbol{\omega}+\boldsymbol{\omega}\cdot\mathbf{J}_3\cdot\mathbf{v}+\boldsymbol{\omega}\cdot\mathbf{J}_4\cdot\boldsymbol{\omega}), \quad \mathbf{K}_1 = \frac{\partial K}{\partial \mathbf{v}}, \quad \mathbf{K}_2 = \frac{\partial K}{\partial \boldsymbol{\omega}}. \quad (4.38)$$

Equation (4.38) imply the following properties of the inertia tensors

$$\mathbf{J}_1 = \mathbf{J}_1^T, \quad \mathbf{J}_2^T = \mathbf{J}_3, \quad \mathbf{J}_4 = \mathbf{J}_4^T.$$

Let us require K to be positive definite. This implies that $\mathbf{J}_1$ and $\mathbf{J}_4$ are positive definite tensors.

We get a particular case of Eq. (4.38) taking $\mathbf{J}_k$ in the form

$$\mathbf{J}_1 = \mathbf{I}, \quad \mathbf{J}_2^T = \mathbf{J}_3 = \mathbf{H}\cdot\mathbf{J}_3^0\cdot\mathbf{H}^T, \quad \mathbf{J}_4 = \mathbf{H}\cdot\mathbf{J}_4^0\cdot\mathbf{H}^T, \qquad (4.39)$$

where $\mathbf{J}_3^0$ and $\mathbf{J}_4^0$ are constant tensors. We will call $\mathbf{J}_3$ and $\mathbf{J}_4$ microinertia tensors in the actual configuration and $\mathbf{J}_3^0$ and $\mathbf{J}_4^0$ microinertia tensors in the reference configuration. Equation (4.39) are completely analogous to Eq. $(B.13)_1$ for the inertia tensor of rigid body.

For further simplification of kinetic equations we can take

$$\mathbf{J}_1 = \mathbf{I}, \quad \mathbf{J}_2 = \mathbf{J}_3 = \mathbf{0}, \quad \mathbf{J}_4 = \rho^{-1} j\,\mathbf{I},$$

that are widely used in the literature, see [35–37]. They correspond to the description of a micropolar particle as a solid homogeneous sphere of finite radius.

4.4 Material Symmetry Group

In general, the form of elastic strain energy density W_κ of the micropolar body depends upon the choice of the reference configuration κ. Particularly important are sets of reference configurations which leave unchanged the form of the energy density. Transformations of the reference configuration under which the energy density remains unchanged are called here *invariant transformations*. In other words invariant transformations are density-preserving deformations and all microrotations of the reference configuration of the micropolar continuum that cannot be experimentally detected. All such invariant transformations constitute the *material symmetry group*. Knowledge of the material symmetry group allows one to precisely define the fluid, the solid, the liquid crystal or the subfluid as well as to introduce notions of isotropic, hemitropic or orthotropic polar-elastic continua among others. A similar approach is used in classical continuum mechanics and in non-linear elasticity in [1, 38–41], as well as in non-linear theories of shells [25]. The material symmetry group of the nonlinear micropolar continuum was first characterized by Eringen and Kafadar [42] and modified by Eremeyev and Pietraszkiewicz [26] by taking into account dependence of W_κ on the microstructure curvature tensor $\mathbf{B}$ as well as different transformation properties of the polar and axial tensors.

Let us introduce another reference configuration κ_* of the micropolar body $\mathscr{B}$, in which the position of point x is given by the vector $\mathbf{R}_*$ relative to the same origin o and its orientation is fixed by three orthonormal directors $\mathbf{D}_{*k}$. Let $\mathbf{P}$, $\det \mathbf{P} \neq 0$, be the deformation gradient transforming $\mathrm{d}\mathbf{R}$ into $\mathrm{d}\mathbf{R}_*$, and $\mathbf{R}$ be the orthogonal tensor transforming $\mathbf{D}_k$ into $\mathbf{D}_{*k}$, so that

$$\mathrm{d}\mathbf{R}_* = \mathbf{P}\cdot \mathrm{d}\mathbf{R}, \quad \mathbf{D}_{*k} = \mathbf{R}\cdot \mathbf{D}_k. \tag{4.40}$$

In what follows all fields associated with deformation relative to the reference configuration κ_* will be marked by the lower index $*$. From (4.40) it follows that

$$\mathbf{F} = \mathbf{F}_* \cdot \mathbf{P}, \quad \mathbf{H} = \mathbf{H}_* \cdot \mathbf{R}. \tag{4.41}$$

The transformations of the strain measures and the microstructure tensors are given by

$$\mathbf{E}_* = \mathbf{R}\cdot\mathbf{E}\cdot\mathbf{P}^{-1} + \mathbf{R}\cdot\mathbf{P}^{-1} - \mathbf{I}, \quad \mathbf{K}_* = (\det\mathbf{R})\mathbf{R}\cdot\mathbf{K}\cdot\mathbf{P}^{-1} + \mathbf{L}, \tag{4.42}$$

$$\mathbf{B}_* = (\det\mathbf{R})\mathbf{R}\cdot\mathbf{B}\cdot\mathbf{P}^{-1} - \mathbf{L}, \tag{4.43}$$

where $\mathbf{L} = \mathbf{R}\cdot\mathbf{Z}\cdot\mathbf{P}^{-1}$, $\mathbf{Z} = -\frac{1}{2}\mathcal{E} : (\mathbf{R}\cdot \mathrm{Grad}\,\mathbf{R}^T)$.

The assumption that the constitutive relation is insensitive to the change of the reference configuration κ into κ_* means that the explicit form of the strain energy densities W_κ and W_* should coincide, that is

$$W_\kappa(\mathbf{E}, \mathbf{K}; \mathbf{R}, \mathbf{B}) = W_\kappa(\mathbf{E}_*, \mathbf{K}_*; \mathbf{R}, \mathbf{B}_*). \tag{4.44}$$

In other words, this means that one may use the same function for the strain energy density independently on the choice of κ or κ_*, but with different expressions for stretch and wryness tensors as well as for the microstructure curvature tensor. In what follows we not always explicitly indicate that all the functions depend also on the position vector $\mathbf{R}$ and W is taken relative to the reference configuration κ.

The relations (4.42)–(4.44) hold locally, i.e. it should be satisfied at any x and $\mathbf{B}$, and the tensors $\mathbf{P}, \mathbf{R}, \mathbf{L}$ are treated as independent here. As a result, the local invariance of W under change of the reference configuration is described by the triple of tensors $(\mathbf{P}, \mathbf{R}, \mathbf{L})$. Using (4.42) and (4.43) we obtain the following invariance requirement for W and introduce the following definition:

Definition 4.1 By the material symmetry group $\mathscr{G}_\kappa$ at x and $\mathbf{B}$ of the micropolar elastic continuum we call all sets of ordered triples of tensors

$$\mathbb{X} = (\mathbf{P} \in Unim, \mathbf{R} \in Orth, \mathbf{L} \in Lin), \tag{4.45}$$

satisfying the relation

$$W(\mathbf{E}, \mathbf{K}; \mathbf{B}) = W\Big[\mathbf{R}\cdot\mathbf{E}\cdot\mathbf{P}^{-1} + \mathbf{R}\cdot\mathbf{P}^{-1} - \mathbf{I},\ (\det\mathbf{R})\mathbf{R}\cdot\mathbf{K}\cdot\mathbf{P}^{-1} + \mathbf{L};\ (\det\mathbf{R})\mathbf{R}\cdot\mathbf{B}\cdot\mathbf{P}^{-1} - \mathbf{L}\Big] \tag{4.46}$$

for any tensors **E**, **K**, **B** in the domain of definition of the function W.

The set $\mathscr{G}_\kappa$ is the group relative to the group operation $\circ$ defined by

$$(\mathbf{P}_1, \mathbf{R}_1, \mathbf{L}_1) \circ (\mathbf{P}_2, \mathbf{R}_2, \mathbf{L}_2) = \Big[\mathbf{P}_1\cdot\mathbf{P}_2,\ \mathbf{R}_1\cdot\mathbf{R}_2,\ \mathbf{L}_1 + (\det\mathbf{R}_1)\mathbf{R}_1\cdot\mathbf{L}_2\cdot\mathbf{P}_1^{-1}\Big].$$

The unit element of $\mathscr{G}_\kappa$ is $\mathbb{I} = (\mathbf{I}, \mathbf{I}, \mathbf{0})$ and the inverse element to $\mathbb{X} \in \mathscr{G}_\kappa$ is given by

$$\mathbb{X}^{-1} \equiv (\mathbf{P}, \mathbf{R}, \mathbf{L})^{-1} = \Big[\mathbf{P}^{-1}, \mathbf{R}^T, -(\det\mathbf{R})\mathbf{R}^T\cdot\mathbf{L}\cdot\mathbf{P}\Big].$$

Since the material symmetry group depends not only on the particle $x \in \mathscr{B}$ but also upon the choice of the reference configuration, let us analyze how the symmetry groups corresponding to different reference configurations are related. Let κ_1 and κ_2 be two different reference configurations, and $\mathscr{G}_1$ and $\mathscr{G}_2$ be the material symmetry groups relative to these reference configurations, respectively. Let now **P** be the non-singular deformation gradient, $\det\mathbf{P} \neq 0$, **R** be the orthogonal tensor associated with transformation $\kappa_1 \to \kappa_2$, as well as $\mathbf{P}^{-1}$ and $\mathbf{R}^T$ be the inverse deformation gradient and the inverse orthogonal tensor associated with the inverse transformation $\kappa_2 \to \kappa_1$, respectively.

The definition (4.45), (4.46) allows one to establish an analogue of the Noll rule given for the classical simple material in [1]. In what follows quantities described in the configurations κ_1 and κ_2 are marked by the respective lower indices 1 and 2. Let the element $\mathbb{X}_1 \equiv (\mathbf{P}_1, \mathbf{R}_1, \mathbf{L}_1) \in \mathscr{G}_1$. Then the element $\mathbb{X}_2 = \mathbb{P}\circ\mathbb{X}_1\circ\mathbb{P}^{-1} \in \mathscr{G}_2$, where $\mathbb{P} \equiv (\mathbf{P}, \mathbf{R}, \mathbf{L})$. Thus the material symmetry group under change of the reference configuration transforms according to the analogue of the *Noll rule*

$$\mathscr{G}_2 = \mathbb{P}\circ\mathscr{G}_1\circ\mathbb{P}^{-1}. \tag{4.47}$$

As in the case of non-polar elastic materials, the property of *isotropy* of the micropolar elastic material is expressed in terms of the orthogonal group.

Definition 4.2 The micropolar elastic continuum is called isotropic at x and **B** if there exists a reference configuration κ, called undistorted, such that the material symmetry group relative to κ contains the group $\mathscr{S}_\kappa$,

$$\mathscr{S}_\kappa \subset \mathscr{G}_\kappa, \quad \mathscr{S}_\kappa \equiv \{(\mathbf{P} = \mathbf{O}, \mathbf{O}, \mathbf{0}) : \quad \mathbf{O} \in Orth\} \tag{4.48}$$

From the physical point of view this definition means that uniform rotations and mirror reflections of the undistorted reference configuration κ cannot be recognized by any experiment.

To define the *micropolar elastic fluid* we apply the requirement that its strain energy density should be insensitive to any change of the reference configuration.

Definition 4.3 The micropolar elastic continuum is called polar micropolar elastic fluid at x and $\mathbf{B}$ if there exists a reference configuration κ, named undistorted, such that the material symmetry group relative to κ is given by

$$\mathscr{G}_\kappa = \mathscr{U}_\kappa \equiv \{(\mathbf{P} \in Unim, \mathbf{R} \in Orth, \mathbf{L} \in Lin)\} \tag{4.49}$$

Hence, the strain energy density of the polar-elastic fluid satisfies the relation

$$\begin{aligned} &W_\kappa(\mathbf{E}, \mathbf{K}; \mathbf{B}) \\ &\quad = W_\kappa \Big[\mathbf{R}\cdot\mathbf{E}\cdot\mathbf{P}^{-1} + \mathbf{R}\cdot\mathbf{P}^{-1} - \mathbf{I},\ (\det\mathbf{R})\mathbf{R}\cdot\mathbf{K}\cdot\mathbf{P}^{-1} \\ &\qquad + \mathbf{L};\ (\det\mathbf{R})\mathbf{R}\cdot\mathbf{B}\cdot\mathbf{P}^{-1} - \mathbf{L}\Big], \quad \forall\mathbf{P}\in Unim, \quad \forall\mathbf{R}\in Orth, \quad \forall\mathbf{L}\in Lin. \end{aligned}$$

From the Noll rule (4.47) it is easy to verify that for the polar-elastic fluid any reference configuration becomes undistorted, similarly as it is for the non-polar elastic fluid, because the symmetry group becomes here to be maximal. Obviously, the micropolar fluid is also isotropic.

The constitutive equation of the polar-elastic fluid are given by, see [22, 23, 26],

$$W = W^\times(\det\mathbf{F}, \mathbf{C}) = \overline{W}(\rho, \mathbf{C}), \tag{4.50}$$

where $\mathbf{C}$ is the microstructure curvature tensor of the deformed configuration χ defined in (4.6). Using Theorem A.4, $W^\times$ can be considered as a function of only six invariants $j_n(\mathbf{C})$, $n = 1, \ldots, 6$, $W = W^\times(\det\mathbf{F}, j_1, j_2, \ldots, j_6)$. The constitutive equations for $\mathbf{T}$ and $\mathbf{M}$ corresponding to (4.50) are

$$\mathbf{T} = -p\mathbf{I} - \mathbf{C}^T\cdot\mathbf{M}, \quad \mathbf{M} = J\frac{\partial\overline{W}}{\partial\mathbf{C}}, \quad p = \rho J\frac{\partial\overline{W}}{\partial\rho}. \tag{4.51}$$

A simple example of the strain energy density is the following quadratic function:

$$\overline{W}(\rho, \mathbf{C}) = \alpha_0(\rho) + \alpha_1 j_1^2 + \alpha_2 j_2 + \alpha_3 j_4, \tag{4.52}$$

where α_i, $i = 1, 2, 3$, are the material constants.

To define the *micropolar elastic solid* we apply the requirement that its material symmetry group consists of the orthogonal tensors.

Definition 4.4 The micropolar elastic continuum is called micropolar elastic solid at x and $\mathbf{B}$ if there exists a reference configuration κ, called undistorted, such that the material symmetry group relative to κ is given by

$$\mathcal{G}_\kappa = \mathcal{R}_\kappa \equiv \{(\mathbf{P} = \mathbf{O}, \mathbf{O}, \mathbf{0}) : \quad \mathbf{O} \in \mathcal{O}_\kappa \subset Orth\}. \tag{4.53}$$

The group $\mathcal{R}_\kappa$ is fully described by a subgroup $\mathcal{O}_\kappa$ of the orthogonal group $Orth$. The invariance requirement of W leads here to finding the subgroup $\mathcal{O}_\kappa$ such that

$$W(\mathbf{E}, \mathbf{K}; \mathbf{B}) = W\left[\mathbf{O} \cdot \mathbf{E} \cdot \mathbf{O}^T, (\det \mathbf{O})\mathbf{O} \cdot \mathbf{K} \cdot \mathbf{O}^T; (\det \mathbf{O})\mathbf{O} \cdot \mathbf{B} \cdot \mathbf{O}^T\right], \quad \forall \mathbf{O} \in \mathcal{O}_\kappa. \tag{4.54}$$

The strain energy density of a polar-elastic continuum may also admit other material symmetry groups, in general. For example, it is possible to construct the material symmetry groups of W in analogy to the symmetry groups used to model liquid crystals in continuum mechanics of simple materials, see [1, 40].

Definition 4.5 The micropolar elastic continuum is called micropolar elastic liquid crystal at x and $\mathbf{B}$ if the material symmetry group $\mathcal{G}_\kappa$ does not coincide with $\mathcal{U}_\kappa$, but there exist elements $\mathbb{X} \in \mathcal{G}_\kappa$, which are not the members of any group constructed using only the orthogonal tensors.

In other words the micropolar polar elastic liquid crystal is the material which is neither fluid nor solid.

4.5 Non-Linear Micropolar Isotropic Solids

Let us consider the micropolar isotropic elastic solids.

Definition 4.6 The micropolar elastic solid is called isotropic at point x and $\mathbf{B}$ if there exists a reference configuration κ, called undistorted, such that the material symmetry group relative to κ takes the form

$$\mathcal{G}_\kappa = \mathcal{S}_\kappa \equiv \{(\mathbf{P} = \mathbf{O}, \mathbf{O}, \mathbf{0}) : \quad \mathbf{O} \in Orth\}. \tag{4.55}$$

This definition means that the strain energy density of the polar-elastic isotropic solid satisfies the relation

$$W(\mathbf{E}, \mathbf{K}; \mathbf{B}) = W\left[\mathbf{O} \cdot \mathbf{E} \cdot \mathbf{O}^T, (\det \mathbf{O})\mathbf{O} \cdot \mathbf{K} \cdot \mathbf{O}^T; (\det \mathbf{O})\mathbf{O} \cdot \mathbf{B} \cdot \mathbf{O}^T\right], \quad \forall \mathbf{O} \in Orth.$$

Scalar-valued isotropic functions of second-order tensors can be expressed by the so-called representation theorems in terms of joint invariants of tensorial arguments [43–45]. The integrity basis for the proper orthogonal group is given by Spencer, see Table 1 in [43] or Table II in [44]. The number of members of the integrity basis of $\mathbf{E}$, $\mathbf{K}$, $\mathbf{B}$ is much larger than the number of components of these tensors. However, there are some polynomial dependencies (syzygies) between elements of the integrity basis of three non-symmetric tensors. For the proper orthogonal group there is no difference in transformations of axial and polar tensors. It is not the case if one considers transformations using the full orthogonal group. Since $\mathbf{K}$ and $\mathbf{B}$ are axial tensors, not all invariants listed in [43, 44] are *absolute invariants* under orthogonal transformations, because some of them change sign under non-proper orthogonal transformations. Following [44], we call such invariants *relative invariants*. Examples of relative invariants are $\operatorname{tr}\mathbf{K}$, $\operatorname{tr}\mathbf{K}^3$, $\operatorname{tr}\mathbf{E}\cdot\mathbf{K}$, $\operatorname{tr}\mathbf{E}\cdot\mathbf{B}$, etc. This gives us the following property of W:

$$W(\mathbf{E},\mathbf{K};\mathbf{B}) = W(\mathbf{E},-\mathbf{K};-\mathbf{B}). \tag{4.56}$$

The full list of 119 absolute and relative invariants for the orthogonal group is presented in [26].

If we neglect the explicit dependence of W on $\mathbf{B}$, or assume that $\mathbf{B} = \mathbf{0}$, then $W = W(\mathbf{E},\mathbf{K})$. The integrity basis of two non-symmentric tensors under the orthogonal group contains 39 members. Following Zheng [45], Ramezani et al. [46] listed these invariants for the non-linear polar-elastic solids and proposed the corresponding constitutive equations. Let us note, however, that not all 39 elements of this integrity basis are functionnally independent. Kafadar and Eringen [3] constructed the functional basis for two non-symmetric tensors taking into account these functional dependencies. According to [3], as the isotropic scalar-valued function of two non-symmetric tensors $\mathbf{E}$ and $\mathbf{K}$, W is expressible in terms of 15 invariants,

$$W = W(I_1, I_2, \ldots, I_{15}), \tag{4.57}$$

where I_k are given by

$$\begin{array}{lll}
I_1 = \operatorname{tr}\mathbf{E}, & I_2 = \operatorname{tr}\mathbf{E}^2, & I_3 = \operatorname{tr}\mathbf{E}^3,\\
I_4 = \operatorname{tr}\mathbf{E}\cdot\mathbf{E}^T, & I_5 = \operatorname{tr}\mathbf{E}^2\cdot\mathbf{E}^T, & I_6 = \operatorname{tr}\mathbf{E}^2\cdot(\mathbf{E}^T)^2,\\
I_7 = \operatorname{tr}\mathbf{E}\cdot\mathbf{K}, & I_8 = \operatorname{tr}\mathbf{E}^2\cdot\mathbf{K}, & I_9 = \operatorname{tr}\mathbf{E}\cdot\mathbf{K}^2,\\
I_{10} = \operatorname{tr}\mathbf{K}, & I_{11} = \operatorname{tr}\mathbf{K}^2, & I_{12} = \operatorname{tr}\mathbf{K}^3,\\
I_{13} = \operatorname{tr}\mathbf{K}\cdot\mathbf{K}^T, & I_{14} = \operatorname{tr}\mathbf{K}^2\cdot\mathbf{K}^T, & I_{15} = \operatorname{tr}\mathbf{K}^2\cdot(\mathbf{K}^T)^2.
\end{array}$$

Taking into account that $W = W(\mathbf{E},\mathbf{K})$ is an even function with respect to $\mathbf{K}$, W becomes also a even function with respect to some invariants

$$\begin{aligned}
&W(I_1, I_2, I_3, I_4, I_5, I_6, I_7, I_8, I_9, I_{10}, I_{11}, I_{12}, I_{13}, I_{14}, I_{15}) \\
&= W(I_1, I_2, I_3, I_4, I_5, I_6, -I_7, -I_8, I_9, -I_{10}, I_{11}, -I_{12}, I_{13}, -I_{14}, I_{15}).
\end{aligned} \tag{4.58}$$

Expanding W a the Taylor series with respect to $\mathbf{E}$ and $\mathbf{K}$, and keeping up to quadratic terms, we obtain with (4.58) the approximate polynomial representation of (4.57)

$$W = w_0 + a_1 I_1 + b_1 I_1^2 + b_2 I_2 + b_3 I_4 + b_4 I_{10}^2 + b_5 I_{11} + b_6 I_{13}, \tag{4.59}$$

where $w_0, a_1, a_2, b_1, \ldots, b_6$ are material parameters.

Let us consider the representation of W which takes the form of a sum of two scalar functions each depending on one strain measure

$$W = W_1(\mathbf{E}) + W_2(\mathbf{K}). \tag{4.60}$$

For example, this form has W in (4.59). Equation (4.60) is also used in [46] in order to generalize the classical neo-Hookean and Mooney-Rivlin models to the micropolar elastic solids. Using the representation theorem A.4 for the isotropic scalar-valued function of one non-symmetric tensor, we obtain the following representation of W:

$$W = \widetilde{W}_1(I_1(\mathbf{E}), \ldots, I_6(\mathbf{E})) + \widetilde{W}_2(I_1(\mathbf{K}), \ldots, I_6(\mathbf{K})), \tag{4.61}$$

where $\widetilde{W}_2$ has the property $\widetilde{W}_2(I_1, I_2, I_3, I_4, I_5, I_6) = \widetilde{W}_2(-I_1, I_2, -I_3, I_4, -I_5, I_6)$.

4.6 Physically Linear Micropolar Solid

Let us consider the strain energy density as a quadratic function of $\mathbf{E}$ and $\mathbf{K}$

$$W = \frac{1}{2}\mathbf{E}\cdot\mathbb{C}\cdot\mathbf{E} + \mathbf{E}\cdot\mathbb{B}\cdot\mathbf{K} + \frac{1}{2}\mathbf{K}\cdot\mathbb{D}\cdot\mathbf{K}, \tag{4.62}$$

where $\mathbb{C}$, $\mathbb{B}$, and $\mathbb{D}$ are the fourth-order tensors of elastic moduli of the micropolar elastic solid. The components of tensors $\mathbb{C}$ and $\mathbb{D}$ have the symmetry properties

$$C_{ijmn} = C_{mnij}, \quad D_{ijmn} = D_{mnij}.$$

With (4.62) the corresponding stress measures $\mathbf{P}_1$ and $\mathbf{P}_2$ take the form

$$\mathbf{P}_1 = \mathbb{C}\cdot\mathbf{E} + \mathbb{B}\cdot\mathbf{K}, \quad \mathbf{P}_2 = \mathbf{E}\cdot\mathbb{B} + \mathbb{D}\cdot\mathbf{K}.$$

The model based on (4.62) can be called the *physically linear micropolar elastic solid*. Since $\mathbf{E}$ and $\mathbf{K}$ are non-symmetric tensors, W contains 171 independent material parameters, in general.

In the case of isotropic solids one can use the representation (A.19). The corresponding tensors $\mathbb{C}$ and $\mathbb{D}$ take the form

$$\mathbb{C} = \lambda\mathbf{I}\otimes\mathbf{I} + \mu\mathbf{i}_a\otimes\mathbf{I}\otimes\mathbf{i}_a + (\mu+\kappa)\mathbf{i}_a\otimes\mathbf{i}_b\otimes\mathbf{i}_a\otimes\mathbf{i}_b, \tag{4.63}$$
$$\mathbb{D} = \beta_1\mathbf{I}\otimes\mathbf{I} + \beta_2\mathbf{i}_a\otimes\mathbf{I}\otimes\mathbf{i}_a + \beta_3\mathbf{i}_a\otimes\mathbf{i}_b\otimes\mathbf{i}_a\otimes\mathbf{i}_b, \tag{4.64}$$

where λ, μ, κ, β_i, $i = 1, 2, 3$, are the independent elastic moduli, while $\mathbb{B} = \mathbb{O}$. Thus, the strain energy density of the physically linear polar-elastic isotropic solid contains only six scalar elastic moduli. W takes form

$$2W = \lambda \mathrm{tr}^2\mathbf{E} + \mu I_2 \mathrm{tr}\,\mathbf{E}^2 + (\mu+\kappa)\mathrm{tr}\,(\mathbf{E}\cdot\mathbf{E}^T) + \beta_1 \mathrm{tr}^2\mathbf{K} + \beta_2 \mathrm{tr}\,\mathbf{K}^2 + \beta_3 \mathrm{tr}\,(\mathbf{K}\cdot\mathbf{K}^T). \quad (4.65)$$

This constitutive relation corresponds to the linear isotropic elastic Cosserat continuum described by (4.67). Within the linear micropolar elasticity the explicit structure of tensors $\mathbb{C}$ and $\mathbb{D}$ is presented by Zheng and Spencer [47] for 14 symmetry groups, see also [48]. Equations of physically linear anisotropic micropolar elastic solid are discussed in [26].

4.7 Linear Micropolar Isotropic Solids

Let us consider the case when the translations and microrotations are very small. Now we assume that $\mathbf{u}$, $\boldsymbol{\theta}$, and their spatial derivatives are infinitesimally small:

$$\|\mathbf{u}\| \ll 1, \quad \|\mathrm{Grad}\,\mathbf{u}\| \ll 1, \quad \|\boldsymbol{\theta}\| \ll 1, \quad \|\mathrm{Grad}\,\boldsymbol{\theta}\| \ll 1,$$

where $\mathbf{u} \triangleq \mathbf{r} - \mathbf{R}$ is the *infinitesimal translation vector*, and $\boldsymbol{\theta} = \varphi\mathbf{e}$ is the *infinitesimal rotation vector* now.

There hold

$$\mathbf{H} \approx \mathbf{I} + \mathbf{I}\times\boldsymbol{\theta}, \quad \mathbf{K} \approx \boldsymbol{\varkappa} \triangleq \mathrm{Grad}\,\boldsymbol{\theta}, \quad \mathbf{E} \approx \boldsymbol{\varepsilon} \triangleq \mathrm{Grad}\,\mathbf{u} - \mathbf{I}\times\boldsymbol{\theta}. \quad (4.66)$$

Definition 4.7 The tensor $\boldsymbol{\varepsilon} \triangleq \mathrm{Grad}\,\mathbf{u} - \mathbf{I}\times\boldsymbol{\theta}$ is called the linear stretch tensor and $\boldsymbol{\varkappa} \triangleq \mathrm{Grad}\,\boldsymbol{\theta}$—the linear wryness tensor.

We should note the introduced linear stretch tensor is different of the strain tensor of linear elasticity $\mathbf{e} \triangleq \frac{1}{2}\left[\mathrm{Grad}\,\mathbf{u} + (\mathrm{Grad}\,\mathbf{u})^T\right]$. The difference for the two theories exists for other quantities, for example, for the tensors of the time rate of strains.

Note in the present linear approximation of the theory, there is no difference between the gradient operators grad and Grad, that is $\mathrm{Grad}\,\mathbf{u} \approx \mathrm{grad}\,\mathbf{u}$, $\mathrm{Grad}\,\boldsymbol{\theta} \approx \mathrm{grad}\,\boldsymbol{\theta}$ as well as there is no difference between the stress tensors $\mathbf{T} \approx \mathbf{T}_\kappa \approx \mathbf{P}_1$ and $\mathbf{M} \approx \mathbf{M}_\kappa \approx \mathbf{P}_2$.

We can rewrite the constitutive equations for isotropic material in the form

$$\mathbf{T} = \frac{\partial W}{\partial \boldsymbol{\varepsilon}}, \quad \mathbf{M} = \frac{\partial W}{\partial \boldsymbol{\varkappa}},$$

where W is the quadratic form of $\boldsymbol{\varepsilon}$, $\boldsymbol{\varkappa}$

$$W = \frac{\lambda}{2}\mathrm{tr}^2\boldsymbol{\varepsilon} + \frac{\mu+\kappa}{2}\mathrm{tr}\,(\boldsymbol{\varepsilon}\cdot\boldsymbol{\varepsilon}^T) + \frac{\mu}{2}\mathrm{tr}\,(\boldsymbol{\varepsilon}\cdot\boldsymbol{\varepsilon}) + \frac{\beta_1}{2}\mathrm{tr}^2\boldsymbol{\varkappa} + \frac{\beta_2}{2}\mathrm{tr}\,(\boldsymbol{\varkappa}\cdot\boldsymbol{\varkappa}) + \frac{\beta_3}{2}\mathrm{tr}\,(\boldsymbol{\varkappa}\cdot\boldsymbol{\varkappa}^T). \quad (4.67)$$

Thus $\mathbf{T}$ and $\mathbf{M}$ are given by linear relations

$$\mathbf{T} = \lambda \mathbf{I} \operatorname{tr} \boldsymbol{\varepsilon} + (\mu + \kappa)\boldsymbol{\varepsilon} + \mu \boldsymbol{\varepsilon}^T, \quad \mathbf{M} = \beta_1 \mathbf{I} \operatorname{tr} \boldsymbol{\varkappa} + \beta_2 \boldsymbol{\varkappa}^T + \beta_3 \boldsymbol{\varkappa}. \tag{4.68}$$

In the framework of the linear theory of the Cosserat continuum, the rotation velocity is given through the material derivative of the microrotation vector: $\boldsymbol{\omega} = \dot{\boldsymbol{\theta}}$.

Quadratic form (4.67) is assumed to be non-negative. This holds if the following inequalities are valid:

$$3\lambda + 2\mu + \kappa \geq 0, \quad 2\mu + \kappa \geq 0, \quad \kappa \geq 0, \tag{4.69}$$

$$3\beta_1 + \beta_2 + \beta_3 \geq 0, \quad \beta_2 + \beta_3 \geq 0, \quad \beta_3 - \beta_2 \geq 0. \tag{4.70}$$

The linear theory of micropolar elasticity is discussed in many works, see for example [35–37] and their reference lists.

In mechanics, it is also known a model of *the Cosserat pseudocontinuum*, also called the *medium with constrained rotations*. By this model, the macrorotation of a body particle that is determined by displacement $\mathbf{u}$, coincides with the microrotations defined by $\boldsymbol{\theta}$. In linear elasticity, the rotation of a body particle is given by formula $-1/2\operatorname{Rot}\mathbf{u}$, cf. [49]. The identity hypothesis for macro- and microrotations brings us to the relation

$$\boldsymbol{\theta} = -\frac{1}{2}\operatorname{Rot}\mathbf{u}. \tag{4.71}$$

Substituting (4.71) to the expression for $\boldsymbol{\varepsilon}$, we get

$$\boldsymbol{\varepsilon} = \frac{1}{2}\left(\operatorname{Grad}\mathbf{u} + (\operatorname{Grad}\mathbf{u})^T\right) = \mathbf{e}.$$

This means that the stretch tensor coincides with the strain tensor of classical linear elasticity. Substituting (4.71) to the expression for $\boldsymbol{\varkappa}$ we can show that $\operatorname{tr}\boldsymbol{\varkappa} = 0$. From (4.68) it follows that for the Cosserat pseudocontinuum the stress and couple tensors possess the properties

$$\mathbf{T} = \mathbf{T}^T, \quad \operatorname{tr}\mathbf{M} = 0.$$

Relation (4.71) also implies that $\boldsymbol{\varkappa}$ can be expressed through the second spatial derivatives of $\mathbf{u}$. Thus the Cosserat pseudocontinuum model can be considered as a particular case of the gradient elasticity theory.

4.8 Constraints

In mechanics, there are various models for the media with constraints. Restrictions on medium deformations are called *constraints*. In nonlinear elasticity, an example of such constraints is the incompressibility of material, it is widely used to describe the deformation of rubber-like materials, see [50]. In nonlinear elasticity, under the incompressibility constraint, there are established some analytical solutions, so-called universal solutions [1]. In this section we consider mechanics of micropolar materials with constraints.

A scalar constraint takes the form

$$l(\mathbf{r}, \mathbf{H}, \operatorname{Grad}\mathbf{r}, \operatorname{Grad}\mathbf{H}) = 0. \tag{4.72}$$

As (4.72) reflects physical properties of the material, it should be material frame-indifferent. Repeating the transformations used to derive W in Sect. 4.4, we get that the frame-indifferent constraint reduces to the following

$$l(\mathbf{E}, \mathbf{K}) = 0. \tag{4.73}$$

Similarly we can show that a vectorial frame-indifferent constraint is of the form

$$\mathbf{H} \cdot \mathbf{l}(\mathbf{E}, \mathbf{K}) = \mathbf{0}. \tag{4.74}$$

The constitutive equations for materials with constraints are deduced with use of Lagrange's multipliers, for example see [51]. Stress tensors $\mathbf{P}_1$ and $\mathbf{P}_2$ are given by formulae

$$\mathbf{P}_1 = \frac{\partial W^*}{\partial \mathbf{E}}, \quad \mathbf{P}_2 = \frac{\partial W^*}{\partial \mathbf{K}},$$

where

$$W^*(\mathbf{E}, \mathbf{K}) = W(\mathbf{E}, \mathbf{K}) + \lambda l(\mathbf{E}, \mathbf{K}) + \boldsymbol{\lambda} \cdot \mathbf{H} \cdot \mathbf{l}(\mathbf{E}, \mathbf{K}),$$

λ and $\boldsymbol{\lambda}$ are the scalar and vectorial Lagrange multipliers related with constrains (4.73) and (4.74), respectively. The both Lagrange's multipliers are frame-indifferent quantities, they arise as a result of constraint reactions. So we have

$$\mathbf{P}_1 = \frac{\partial W}{\partial \mathbf{E}} + \lambda \frac{\partial l}{\partial \mathbf{E}} + \boldsymbol{\lambda} \cdot \mathbf{H} \cdot \frac{\partial \mathbf{l}}{\partial \mathbf{E}}, \quad \mathbf{P}_2 = \frac{\partial W}{\partial \mathbf{K}} + \lambda \frac{\partial l}{\partial \mathbf{K}} + \boldsymbol{\lambda} \cdot \mathbf{H} \cdot \frac{\partial \mathbf{l}}{\partial \mathbf{K}}.$$

Note that instead of $\boldsymbol{\lambda}$ we can use a non-indifferent vector $\tilde{\boldsymbol{\lambda}} = \boldsymbol{\lambda} \cdot \mathbf{H}$.

An example of scalar constraints is the *incompressibility* condition, $\det \mathbf{F} = 1$. In terms of strain measures, it takes the form

$$l \equiv \det(\mathbf{E} + \mathbf{I}) - 1 = 0. \tag{4.75}$$

For a micropolar body, the incompressibility condition was used in [31, 32, 46].

Using a linear model of the Cosserat pseudocontinuum we will present an example of vectorial constraint. In the polar decomposition of deformation gradient

$$\mathbf{F} = \mathbf{O} \cdot \mathbf{V},$$

we identify microrotation $\mathbf{H}$ with the rotation tensor $\mathbf{O}$. We remind that $\mathbf{V}$ is a positive definite tensor that is $\mathbf{V} = (\mathbf{F}^T \cdot \mathbf{F})^{1/2}$ and $\mathbf{O}$ is an orthogonal tensor. So $\mathbf{H} = \mathbf{O}$ and

$$\mathbf{E} = \mathbf{V} - \mathbf{I}.$$

As in the linear case, it is seen that $\mathbf{E}$ is symmetric. Instead of tensorial relations $\mathbf{H} = \mathbf{O}$ and $\mathbf{E} = \mathbf{E}^T$ we can use an equivalent vectorial constraint

$$\mathbf{l} \equiv \mathbf{E}_\times = \mathbf{0}. \tag{4.76}$$

Condition (4.76) was applied in [31, 32].

Using $(4.66)_3$ for infinitesimal deformations, we can see that (4.76) takes the form of (4.71) of the Cosserat linear pseudocontinuum model.

Note that for finite deformations, constraint (4.76) affects the stress tensor definition but does not restrict the form of the couple stress tensor. However, in the Cosserat linear pseudocontinuum theory, constraint (4.71) involves that $\operatorname{tr}\mathbf{M} = 0$. As its consequence in this theory we cannot assign three independent couple boundary conditions.

For constraints (4.75) and (4.76), $\mathbf{P}_1$ takes the form

$$\mathbf{P}_1 = \frac{\partial W}{\partial \mathbf{E}} + \lambda \mathbf{H}^T \cdot \mathbf{F}^{-T} - \tilde{\boldsymbol{\lambda}} \times \mathbf{I}.$$

To derive this, we should use the differentiation formulae

$$(\det \mathbf{X})_{,\mathbf{X}} = (\det \mathbf{X})\mathbf{X}^{-T}, \quad (\mathbf{X}_\times)_{,\mathbf{X}} = \boldsymbol{\mathcal{E}}.$$

As an example of constitutive equations with constraints we present a generalization of Varga's incompressible material model [52] to micropolar elasticity with rotation constraints:

$$W = \mu \operatorname{tr}\mathbf{E} + \frac{\beta_1}{2}\operatorname{tr}^2\mathbf{K} + \frac{\beta_2}{2}\operatorname{tr}(\mathbf{K}\cdot\mathbf{K}^T) + \frac{\beta_3}{2}\operatorname{tr}(\mathbf{K}\cdot\mathbf{K}). \tag{4.77}$$

For model (4.77), tensors $\mathbf{P}_1$ and $\mathbf{P}_2$ are

$$\mathbf{P}_1 = \mu\mathbf{I} + \lambda\mathbf{H}^T \cdot \mathbf{F}^{-T} - \tilde{\boldsymbol{\lambda}} \times \mathbf{I}, \quad \mathbf{P}_2 = \beta_1 \mathbf{I}\operatorname{tr}\mathbf{K} + \beta_2\mathbf{K} + \beta_3\mathbf{K}^T. \tag{4.78}$$

In a similar way, well known models of incompressible elastic materials by Mooney–Rivlin, Ogden, etc., can be extended to the Cosserat nonlinear pseudocontinuum.

It is worth of noting that the rigid body motion can be considered as a particular case of the motion of a micropolar body with the following constraints: $\mathbf{E} = \mathbf{K} = \mathbf{0}$. In this case, stress tensor $\mathbf{P}_1$ and couple tensor $\mathbf{P}_2$ are Lagrange "multipliers" that correspond to the constraints and do not depend on the deformation. The Euler laws of motion take the form of (B.20).

4.9 Constitutive Inequalities

In nonlinear elasticity there are known so-called constitutive restrictions. They are the strong ellipticity condition, Hadamard inequality, generalized Coleman–Noll condition (GCN-condition), and some others [1, 39, 53]. Each of them plays a role in nonlinear elasticity. They express mathematically precise and physically intuitive restrictions for constitutive equations of elastic bodies. In particular, the GCN condition proposed by Coleman and Noll asserts that "the transformation from deformation gradient to first Piola–Kirchhoff stress tensor shall be monotone with respect to pairs of deformations differing from one another by a pure stretch" (see, [1]).

4.9.1 Constitutive Restrictions in Linear Micropolar Elasticity

The simplest example of the constitutive restrictions gives the linear micropolar elasticity. Suppose that the strain energy density of the linear elastic micropolar solids $W(\boldsymbol{\varepsilon}, \boldsymbol{\varkappa})$ is positive definite

$$W(\boldsymbol{\varepsilon}, \boldsymbol{\varkappa}) > 0, \quad \forall \boldsymbol{\varepsilon}, \boldsymbol{\varkappa} \neq \mathbf{0}$$

For an isotropic micropolar elastic solid, W takes the form (4.67). Positivity of W with respect to $\boldsymbol{\varepsilon}$ and $\boldsymbol{\varkappa}$ is equivalent to the strict inequalities (4.69) and (4.70). If the inequalities fail (4.69) and (4.70) this leads to a number of pathological mathematical and physical consequences. For example, boundary value problems of linear micropolar elasticity can have more than one solutions or can have no solution for some loads.

For finite strains, the positive definiteness of the strain energy density $W(\mathbf{E}, \mathbf{K})$ with respect to the strain measures is not enough for the desired properties of boundary value problems. As in the case of non-linear elasticity, one should introduce some additional restrictions.

4.9.2 Coleman–Noll Inequality for Micropolar Continuum

The Coleman–Noll constitutive inequality is one of well-known in nonlinear elasticity [1, 39, 53]. Its differential form, a so-called GCN-condition, expresses the property that the work of the stress rate cannot be negative or the property that the first Piola–Kirchhoff stress tensor shall be monotone with respect to pairs of deformations differing from one another by a pure stretch.

Generalizing the classical GCN-condition let us assume that

$$\dot{\mathbf{P}}_1 \cdot \dot{\mathbf{E}} + \dot{\mathbf{P}}_2 \cdot \dot{\mathbf{K}} > 0, \quad \forall \dot{\mathbf{E}} \neq \mathbf{0}, \dot{\mathbf{K}} \neq \mathbf{0}. \tag{4.79}$$

The strain rates are given by

$$\dot{\mathbf{E}} = \mathbf{H}^T \cdot (\mathrm{Grad}\,\mathbf{v} - \boldsymbol{\omega} \times \mathbf{F}) = \mathbf{H}^T \cdot \boldsymbol{\varepsilon} \cdot \mathbf{F}, \quad \dot{\mathbf{K}} = \mathbf{H}^T \cdot \mathrm{Grad}\,\boldsymbol{\omega} = \mathbf{H}^T \cdot \boldsymbol{\varkappa} \cdot \mathbf{F}, \tag{4.80}$$

where $\boldsymbol{\varepsilon}$ and $\boldsymbol{\varkappa}$ are the strain rates given by

$$\boldsymbol{\varepsilon} \triangleq \mathrm{grad}\,\mathbf{v} - \boldsymbol{\omega} \times \mathbf{I}, \quad \boldsymbol{\varkappa} \triangleq \mathrm{grad}\,\boldsymbol{\omega}. \tag{4.81}$$

Taking into account (4.30) and (4.80), (4.81), we rewrite (4.79) in the equivalent form

$$\left.\frac{\mathrm{d}^2}{\mathrm{d}\tau^2} W(\mathbf{E} + \tau\boldsymbol{\varepsilon}, \mathbf{K} + \tau\boldsymbol{\kappa})\right|_{\tau=0} > 0 \ \ \forall\, \boldsymbol{\varepsilon} \neq \mathbf{0}, \quad \boldsymbol{\kappa} \neq \mathbf{0}. \tag{4.82}$$

Indeed, from $(4.30)_{1,2}$ it follows

$$\dot{\mathbf{P}}_1 = \frac{\partial^2 W}{\partial \mathbf{E} \partial \mathbf{E}} \cdot \dot{\mathbf{E}} + \frac{\partial^2 W}{\partial \mathbf{E} \partial \mathbf{K}} \cdot \dot{\mathbf{K}}, \quad \dot{\mathbf{P}}_2 = \frac{\partial^2 W}{\partial \mathbf{K} \partial \mathbf{E}} \cdot \dot{\mathbf{E}} + \frac{\partial^2 W}{\partial \mathbf{K} \partial \mathbf{K}} \cdot \dot{\mathbf{K}}, \tag{4.83}$$

and (4.79) takes the following form:

$$\dot{\mathbf{E}} \cdot \frac{\partial^2 W}{\partial \mathbf{E} \partial \mathbf{E}} \cdot \dot{\mathbf{E}} + 2\dot{\mathbf{E}} \cdot \frac{\partial^2 W}{\partial \mathbf{E} \partial \mathbf{K}} \cdot \dot{\mathbf{K}} + \dot{\mathbf{K}} \cdot \frac{\partial^2 W}{\partial \mathbf{K} \partial \mathbf{K}} \cdot \dot{\mathbf{K}} > 0, \quad \forall \dot{\mathbf{E}}, \dot{\mathbf{K}} \neq \mathbf{0}. \tag{4.84}$$

On the other hand the inequality (4.82) can be transformed to

$$\boldsymbol{\varepsilon} \cdot \frac{\partial^2 W}{\partial \mathbf{E} \partial \mathbf{E}} \cdot \boldsymbol{\varepsilon} + 2\boldsymbol{\varepsilon} \cdot \frac{\partial^2 W}{\partial \mathbf{E} \partial \mathbf{K}} \cdot \boldsymbol{\varkappa} + \boldsymbol{\varkappa} \cdot \frac{\partial^2 W}{\partial \mathbf{K} \partial \mathbf{K}} \cdot \boldsymbol{\varkappa} > 0, \quad \forall \boldsymbol{\varepsilon}, \boldsymbol{\varkappa} \neq \mathbf{0}. \tag{4.85}$$

Taking into account (4.81) and the fact that $\dot{\mathbf{E}}$, $\dot{\mathbf{K}}$ and $\boldsymbol{\varepsilon}$, $\boldsymbol{\varkappa}$ are arbitrary non-zero second-order tensors, it is obvious that (4.84) and (4.85) are equivalent to each other.

The conditions (4.82) or (4.79) are analogues of the *Coleman–Noll inequality* in the 3D elasticity. Note that the inequality (4.82) satisfies the principle of material frame-indifference. This means that (4.82) can be chosen as a constitutive inequality for micropolar elastic continuum.

4.9.3 Strong Ellipticity and Hadamard Inequality

The *strong ellipticity* condition is one of well known constitutive restrictions in nonlinear elasticity, cf. [1]. Mathematically, it expresses a precise and physically intuitive restriction for the constitutive equations of elastic materials. In this theory, ellipticity of the equilibrium equations is considered in a number of papers and books, cf. [39, 51, 54–57].

In general, the strong ellipticity of the equations is a property of the material in some area of deformation. For some deformations it can be fulfilled whereas for others it does not. For a number of real media that are modeled by the Cosserat continuum, arising of discontinuous solution is physically possible. For example, such media are soils, granular and porous media. For some other materials such discontinuities are physically impossible. Thus, the strong ellipticity condition allows us "to sort" admissible types of the constitutive equations as well as to determine "dangerous" deformations. Besides, the condition of strong ellipticity is a criterion to find "dangerous" domains in a body.

To formulate the strong ellipticity condition of the equilibrium equations (3.42) we use the general results of the partial differential equations theory (PDE) [58–60]. The linearized equilibrium equations are

$$\operatorname{Div}\dot{\mathbf{T}}_\kappa + \rho_0\dot{\mathbf{f}} = \mathbf{0}, \quad \operatorname{Div}\dot{\mathbf{M}}_\kappa - \left(\dot{\mathbf{T}}_\kappa \cdot \mathbf{F}^T + \mathbf{T}_\kappa \cdot \dot{\mathbf{F}}^T\right)_\times + \rho_0\dot{\mathbf{m}} = \mathbf{0}, \tag{4.86}$$

where the operation $\dot{(\ldots)}$ is similar to the derivative with respect to time. The following formulae hold

$$\dot{\mathbf{T}}_\kappa = \dot{\mathbf{H}} \cdot \mathbf{P}_1 + \mathbf{H} \cdot \dot{\mathbf{P}}_1, \quad \dot{\mathbf{M}}_\kappa = \dot{\mathbf{H}} \cdot \mathbf{P}_2 + \mathbf{H} \cdot \dot{\mathbf{P}}_2, \quad \dot{\mathbf{H}} = \boldsymbol{\omega} \times \mathbf{H}, \quad \dot{\mathbf{F}} = \operatorname{Grad}\mathbf{v},$$

where $\dot{\mathbf{P}}_1$ and $\dot{\mathbf{P}}_2$ are given by Eqs. (4.83) and (4.80). In what follows we restrict ourselves by the dead loading, that is when $\dot{\mathbf{f}} = \mathbf{0}$, $\dot{\mathbf{m}} = \mathbf{0}$.

Equation (4.86) constitute a system of linear PDE of second order with respect to $\mathbf{v}$ and $\boldsymbol{\omega}$. To formulate the strong ellipticity condition we construct the *principal symbol* of (4.86). In other words we select in (4.86) the terms containing the second derivatives of $\mathbf{v}$ and $\boldsymbol{\omega}$. These are

$$\operatorname{Div}\left\{\mathbf{H} \cdot \left[\frac{\partial^2 W}{\partial\mathbf{E}\partial\mathbf{E}} \cdot \left(\mathbf{H}^T \cdot \operatorname{Grad}\mathbf{v}\right) + \frac{\partial^2 W}{\partial\mathbf{E}\partial\mathbf{K}} \cdot \left(\mathbf{H}^T \cdot \operatorname{Grad}\boldsymbol{\omega}\right)\right]\right\},$$
$$\operatorname{Div}\left\{\mathbf{H} \cdot \left[\frac{\partial^2 W}{\partial\mathbf{K}\partial\mathbf{E}} \cdot \left(\mathbf{H}^T \cdot \operatorname{Grad}\mathbf{v}\right) + \frac{\partial^2 W}{\partial\mathbf{K}\partial\mathbf{K}} \cdot \left(\mathbf{H}^T \cdot \operatorname{Grad}\boldsymbol{\omega}\right)\right]\right\}.$$

Following a formal procedure [59], we replace the differential operators Grad and Div by algebraic operations with the unit vector $\mathbf{N}$ and vector fields $\mathbf{v}$ and $\boldsymbol{\omega}$ by vectors $\mathbf{a}$ and $\mathbf{b}$, respectively. This means that we replace $\operatorname{Div}(\ldots)$ by $(\ldots) \cdot \mathbf{N}$, $\operatorname{Grad}\mathbf{u}$ by $\mathbf{a} \otimes \mathbf{N}$, and $\operatorname{Grad}\boldsymbol{\omega}$ by $\mathbf{b} \otimes \mathbf{N}$. Hence we obtain the algebraic expressions

$$\mathbf{H} \cdot \left[\frac{\partial^2 W}{\partial\mathbf{E}\partial\mathbf{E}} \cdot \left(\mathbf{H}^T \cdot \mathbf{a} \otimes \mathbf{N}\right) + \frac{\partial^2 W}{\partial\mathbf{E}\partial\mathbf{K}} \cdot \left(\mathbf{H}^T \cdot \mathbf{b} \otimes \mathbf{N}\right)\right] \cdot \mathbf{N},$$
$$\mathbf{H} \cdot \left[\frac{\partial^2 W}{\partial\mathbf{K}\partial\mathbf{E}} \cdot \left(\mathbf{H}^T \cdot \mathbf{a} \otimes \mathbf{N}\right) + \frac{\partial^2 W}{\partial\mathbf{K}\partial\mathbf{K}} \cdot \left(\mathbf{H}^T \cdot \mathbf{b} \otimes \mathbf{N}\right)\right] \cdot \mathbf{N}.$$

Multiplying the first equation by vector $\mathbf{a}$, the second equation by $\mathbf{b}$ and adding the results we derive the *strong ellipticity* condition of Eq. (3.42):

$$\mathbf{a}\cdot\left\{\mathbf{H}\cdot\left[\frac{\partial^2 W}{\partial\mathbf{E}\partial\mathbf{E}}\cdot\left(\mathbf{H}^T\cdot\mathbf{a}\otimes\mathbf{N}\right)+\frac{\partial^2 W}{\partial\mathbf{E}\partial\mathbf{K}}\cdot\left(\mathbf{H}^T\cdot\mathbf{b}\otimes\mathbf{N}\right)\right]\right\}\cdot\mathbf{N}$$
$$+\mathbf{b}\cdot\left\{\mathbf{H}\cdot\left[\frac{\partial^2 W}{\partial\mathbf{K}\partial\mathbf{E}}\cdot\left(\mathbf{H}^T\cdot\mathbf{a}\otimes\mathbf{N}\right)+\frac{\partial^2 W}{\partial\mathbf{K}\partial\mathbf{K}}\cdot\left(\mathbf{H}^T\cdot\mathbf{b}\otimes\mathbf{N}\right)\right]\right\}\cdot\mathbf{N}>0,\quad\forall\mathbf{a},\mathbf{b}\neq\mathbf{0}.$$

We transform the latter inequality as follows

$$\left(\mathbf{H}^T\cdot\mathbf{a}\otimes\mathbf{N}\right)\cdot\frac{\partial^2 W}{\partial\mathbf{E}\partial\mathbf{E}}\cdot\left(\mathbf{H}^T\cdot\mathbf{a}\otimes\mathbf{N}\right)+2\left(\mathbf{H}^T\cdot\mathbf{a}\otimes\mathbf{N}\right)\cdot\frac{\partial^2 W}{\partial\mathbf{E}\partial\mathbf{K}}\cdot\left(\mathbf{H}^T\cdot\mathbf{b}\otimes\mathbf{N}\right)$$
$$+\left(\mathbf{H}^T\cdot\mathbf{b}\otimes\mathbf{N}\right)\frac{\partial^2 W}{\partial\mathbf{K}\partial\mathbf{K}}\cdot\left(\mathbf{H}^T\cdot\mathbf{b}\otimes\mathbf{N}\right)>0,\quad\forall\mathbf{a},\mathbf{b}\neq\mathbf{0}.$$

Using matrix notations we rewrite the inequality in a compact form

$$\boldsymbol{\xi}\cdot\mathbb{Q}(\mathbf{N})\cdot\boldsymbol{\xi}>0,\quad\forall\,\mathbf{N}\neq\mathbf{0},\quad\forall\,\boldsymbol{\xi}\in\mathbb{R}^6,\quad\boldsymbol{\xi}\neq\mathbf{0},\tag{4.87}$$

where $\boldsymbol{\xi}=(\mathbf{a}',\mathbf{b}')\in\mathbb{R}^6$, $\mathbf{a}'=\mathbf{a}\cdot\mathbf{H}$, $\mathbf{b}'=\mathbf{b}\cdot\mathbf{H}$, and matrix $\mathbb{Q}(\mathbf{N})$ is

$$\mathbb{Q}(\mathbf{N})\triangleq\begin{bmatrix}\dfrac{\partial^2 W}{\partial\mathbf{E}\partial\mathbf{E}}\{\mathbf{N}\} & \dfrac{\partial^2 W}{\partial\mathbf{E}\partial\mathbf{K}}\{\mathbf{N}\}\\[2ex] \dfrac{\partial^2 W}{\partial\mathbf{K}\partial\mathbf{E}}\{\mathbf{N}\} & \dfrac{\partial^2 W}{\partial\mathbf{K}\partial\mathbf{K}}\{\mathbf{N}\}\end{bmatrix},$$

where for arbitrary forth-order tensor $\mathbb{G}$ and vector $\mathbf{N}$ that are represented in a Cartesian basis $\mathbf{i}_k$ ($k=1,2,3$), we have used the notation

$$\mathbb{G}\{\mathbf{N}\}\equiv G_{klmn}N_lN_n\mathbf{i}_k\otimes\mathbf{i}_m.$$

For example, if $\mathbb{G}=\mathbf{c}\otimes\mathbf{d}\otimes\mathbf{e}\otimes\mathbf{f}$ then $\mathbb{G}\{\mathbf{N}\}=(\mathbf{d}\cdot\mathbf{N})(\mathbf{f}\cdot\mathbf{N})\mathbf{c}\otimes\mathbf{e}$.

Inequality (4.87) is the *strong ellipticity condition* of the equilibrium equations (3.42) for the micropolar elastic solids. A weak form of inequality (4.87) is an analogue of the *Hadamard inequality* , that is

$$\boldsymbol{\xi}\cdot\mathbb{Q}(\mathbf{N})\cdot\boldsymbol{\xi}\geq 0,\quad\forall\mathbf{N}\neq\mathbf{0},\quad\forall\boldsymbol{\xi}\in\mathbb{R}^6,\quad\boldsymbol{\xi}\neq\mathbf{0},\tag{4.88}$$

$\mathbb{Q}$ plays a role of the *acoustic tensor* in the micropolar elasticity.

The strong ellipticity condition and the Hadamard inequality are also examples of constitutive restrictions of the constitutive equations of the micropolar elastic solids under finite deformations. As for the theory of non-polar materials, a failure in inequality (4.88) can lead to the existence of non-smooth solutions to equilibrium equations (3.42).

As the Coleman–Noll inequality the strong ellipticity condition can be written in the equivalent form

$$\frac{d^2}{d\tau^2} W\left(\mathbf{E} + \tau \mathbf{a}' \otimes \mathbf{N}, \mathbf{K} + \tau \mathbf{b}' \otimes \mathbf{N}\right)\Big|_{\tau=0} > 0 \quad \forall\, \mathbf{N},\ \mathbf{a}',\ \mathbf{b}' \neq \mathbf{0}. \tag{4.89}$$

Comparing (4.82) and (4.89) we prove the following theorem:

Theorem 4.1 *For the non-liner micropolar elastic solid the Coleman–Noll inequality implies the strong ellipticity condition.*

Proof Indeed, inequality (4.82) holds for any tensors $\boldsymbol{\varepsilon}$ and $\boldsymbol{\kappa}$. Unlike to 3D non-linear elasticity here $\boldsymbol{\varepsilon}$ and $\boldsymbol{\kappa}$ may be nonsymmetric tensors. Substituting relations $\boldsymbol{\varepsilon} = \mathbf{a}' \otimes \mathbf{N}$ and $\boldsymbol{\kappa} = \mathbf{b}' \otimes \mathbf{N}$ into (4.82), we immediately obtain (4.89). □

Thus, for the micropolar elastic continuum the strong ellipticity condition is a particular case of the Coleman–Noll inequality. This is an essential difference between the micropolar elasticity and the non-linear elasticity of non-polar elastic materials [1, 39]: in the latter these two properties are independent in the sense that neither of them implies the other.

Let us consider the constitutive relation (4.60). Then the Hadamard inequality (4.87) is equivalent to two independent inequalities

$$\mathbf{a}\cdot\frac{\partial^2 W_1}{\partial \mathbf{E} \partial \mathbf{E}}\{\mathbf{N}\}\cdot\mathbf{a} > 0, \quad \mathbf{b}\cdot\frac{\partial^2 W_2}{\partial \mathbf{K} \partial \mathbf{K}}\{\mathbf{N}\}\cdot\mathbf{b} > 0, \quad \forall\, \mathbf{N},\ \mathbf{a},\ \mathbf{b} \neq \mathbf{0}.$$

As an example, let us consider consequences of (4.87) for the physically linear isotropic micropolar solid (4.65). In this case the second-order tensors $\dfrac{\partial^2 W_1}{\partial \mathbf{E} \partial \mathbf{E}}\{\mathbf{N}\}$ and $\dfrac{\partial^2 W_2}{\partial \mathbf{K} \partial \mathbf{K}}\{\mathbf{N}\}$ have the form

$$\begin{aligned} \frac{\partial^2 W_1}{\partial \mathbf{E} \partial \mathbf{E}}\{\mathbf{N}\} &= (\mu + \kappa)\,\mathbf{I} + (\lambda + \mu)\mathbf{N} \otimes \mathbf{N}, \\ \frac{\partial^2 W_2}{\partial \mathbf{K} \partial \mathbf{K}}\{\mathbf{N}\} &= \beta_3 \mathbf{I} + (\beta_1 + \beta_2)\mathbf{N} \otimes \mathbf{N}. \end{aligned} \tag{4.90}$$

Using (4.90) we show that the inequality (4.87) is equivalent to the following conditions

$$2\mu + \kappa + \lambda > 0, \quad \mu + \kappa > 0, \quad \beta_1 + \beta_2 + \beta_3 > 0, \quad \beta_3 > 0. \tag{4.91}$$

Indeed, from (4.90) it follows

$$\mathbf{a} \cdot \frac{\partial^2 W_1}{\partial \mathbf{E} \partial \mathbf{E}}\{\mathbf{N}\} \cdot \mathbf{a} = (\lambda + \mu)(\mathbf{a} \cdot \mathbf{N})^2 + (\mu + \kappa)\,\mathbf{a} \cdot \mathbf{a} > 0.$$

Decomposing $\mathbf{a}$ as the sum $\mathbf{a} = a_N \mathbf{N} + \mathbf{a}_\perp$, where $a_N = \mathbf{a} \cdot \mathbf{N}$ and $\mathbf{a}_\perp = \mathbf{a} - a_N \mathbf{N}$ we obtain

$$\mathbf{a} \cdot \frac{\partial^2 W_1}{\partial \mathbf{E} \partial \mathbf{E}}\{\mathbf{N}\} \cdot \mathbf{a} = (\lambda + 2\mu + \kappa)\, a_N^2 + (\mu + \kappa)\, \mathbf{a}_\perp \cdot \mathbf{a}_\perp > 0.$$

The latter inequality implies(4.91)$_{1,2}$. The inequalities (4.91)$_{3,4}$ can be proven in a similar manner.

Thus for the strain energy density (4.65), the strong ellipticity condition reduces to simple inequalities (4.91). They are expressed in terms of the elastic moduli and do not depend on strains. For simple non-linear solids the analogue of (4.65) is the semi-linear material, see [49, 51]. For the latter, the strong elipticity condition depends on strains as well as the elastic moduli. When the ellipticity condition breaks down, singular solutions arise—these may correspond to infinite rotations, for example, [51]. Hence it may be simpler to check the strong ellipticity condition for a micropolar material than for a simple nonlinearly elastic material.

In the case of small strains, the strain energy density (4.65) takes the form of (4.67). Following from the condition of positive semidefiniteness of W, the inequalities (4.69) and (4.70) imply (4.89) but not conversely. Obviously, the strong ellipticity conditions are less restrictive than the requirement of positive semidefiniteness of the energy under infinitesimal deformations.

Eremeyev and Zubov [31] proved that the infinitesimal stability of micropolar body implies the strong ellipticity condition. The proof follows from the analysis of the second variation of the total energy functional of the micropolar elastic body. Unlike to the 3D non-linear elasticity, for the micropolar elasticity the fulfillment of the strong ellipticity condition does not imply the infinitesimal stability of a homogeneous deformation of a micropolar body with Dirichlet boundary conditions.

4.9.4 Ordinary Ellipticity

Let us introduce another type of ellipticity that is the ordinary ellipticity. We consider the solution of (3.42) which have the *weak discontinuity* on a singular surface S. This means that we assume the existence of a singular surface S on which there happen a jump in the values of second spatial derivatives of the position vector $\mathbf{r}$ or microrotation tensor $\mathbf{H}$.

From (3.42) it follows

$$[\![\mathrm{Div}\,\mathbf{T}_\kappa]\!] = \mathbf{0}, \quad [\![\mathrm{Div}\,\mathbf{M}_\kappa]\!] = \mathbf{0}. \tag{4.92}$$

Repeating the transformations of the previous section, we transform (4.92) to

$$\begin{bmatrix} \dfrac{\partial^2 W}{\partial \mathbf{E} \partial \mathbf{E}}\{\mathbf{N}\} & \dfrac{\partial^2 W}{\partial \mathbf{E} \partial \mathbf{K}}\{\mathbf{N}\} \\[2ex] \dfrac{\partial^2 W}{\partial \mathbf{K} \partial \mathbf{E}}\{\mathbf{N}\} & \dfrac{\partial^2 W}{\partial \mathbf{K} \partial \mathbf{L}}\{\mathbf{N}\} \end{bmatrix} \cdot \begin{bmatrix} \mathbf{a}' \\[2ex] \mathbf{b}' \end{bmatrix} = \mathbf{0}$$

We rewrite this in a compact form

$$\mathbf{Q}(\mathbf{N})\cdot\boldsymbol{\xi} = \mathbf{0}, \quad \boldsymbol{\xi} = (\mathbf{a}', \mathbf{b}') \in \mathbb{R}^6. \tag{4.93}$$

The non-zero solutions of Eq. (4.93), that are solutions having weak discontinuity, exist if the determinant of the matrix $\mathbf{A}(\mathbf{N})$ is zero. If

$$\det \mathbf{Q}(\mathbf{N}) \neq 0, \quad \forall \mathbf{N}, \tag{4.94}$$

the weak discontinuities are impossible.

Condition (4.94) is the *ellipticity condition* of the equilibrium equations of micropolar elastic continuum named also the condition of the *Petrovsky ellipticity* or of the *ordinary ellipticity*. The condition follows from the general definition of ellipticity in the partial differential equations theory [60–62]. Obviously, the inequalities (4.94) are weaker than the strong ellipticity condition (4.87).

For the constitutive relation (4.60), condition (4.94) splits into two conditions

$$\det \frac{\partial^2 W_1}{\partial \mathbf{E} \partial \mathbf{E}}\{\mathbf{N}\} \neq 0, \quad \det \frac{\partial^2 W_2}{\partial \mathbf{K} \partial \mathbf{K}}\{\mathbf{N}\} \neq 0. \tag{4.95}$$

As an example, we consider conditions (4.95) for the constitutive relations of a physically linear micropolar solid (4.65). Using Eq. (4.90), we reduce (4.95) to the inequalities

$$\mu + \kappa \neq 0, \quad 2\mu + \lambda + \kappa \neq 0, \quad \beta_3 \neq 0, \quad \beta_1 + \beta_2 + \beta_3 \neq 0.$$

Let us note that for the micropolar elastic solids with constraints we have mixed order systems of PDE. In this case one may use a more general definition of ellipticity such as the ellipticity in the sense of Douglis–Nirenberg [61].

4.10 Micropolar Fluid

Although the basic content of the book is devoted to the micropolar elastic solids in this section we briefly recall the constitutive equations of viscous micropolar fluids. The model was proposed by Aero et al. [63] and Eringen [64], see [65–67]. Some generalizations of the model are presented in [22, 23, 68–70]. For the sake of simplicity we restrict ourselves by the equations of the *linear viscous incompressible micropolar fluid*. The constitutive equations for the stress and couple stress tensors are given by

$$\mathbf{T} = -p\mathbf{I} + \mu_1 \boldsymbol{\varepsilon} + \mu_2 \boldsymbol{\varepsilon}^T, \quad \mathbf{M} = \nu_1 (\operatorname{tr} \boldsymbol{\varkappa}) \mathbf{I} + \nu_2 \boldsymbol{\varkappa} + \nu_3 \boldsymbol{\varkappa}^T, \tag{4.96}$$

where p is the pressure, the strain rates $\boldsymbol{\varepsilon}$ and $\boldsymbol{\varkappa}$ are defined by Eq. (4.81), and μ_1, μ_2, ν_1, ν_2, ν_3 are the viscosities. The incompressibility condition of the fluid is

$$\operatorname{tr} \boldsymbol{\varepsilon} \equiv \operatorname{div} \mathbf{v} = 0. \tag{4.97}$$

Equation (4.96) satisfy the material frame-indifference principle and represent the general possible linear dependence of $\mathbf{T}$ and $\mathbf{M}$ on $\boldsymbol{\varepsilon}$ and $\boldsymbol{\varkappa}$.

Assuming that the dissipation should be positive we obtain the following inequality

$$\mathbf{T} \cdot \boldsymbol{\varepsilon} + \mathbf{M} \cdot \boldsymbol{\varkappa} \geq 0, \quad \forall \boldsymbol{\varepsilon}, \boldsymbol{\varkappa} \neq \mathbf{0}. \tag{4.98}$$

This implies the constitutive inequalities for the viscosity parameters

$$\mu_1 + \mu_2 \geq 0, \quad \mu_1 - \mu_2 \geq 0, \quad \nu_1 \geq 0, \quad \nu_2 + \nu_3 \geq 0, \quad \nu_2 - \nu_3 \geq 0.$$

4.11 Some Sources of Constitutive Equations for Micropolar Materials

For any mechanical theory, the key point is the formulation of constitutive equations. The theory should propose the general form of constitutive equations and the ways how to get their numerical realization for certain materials. For a Cosserat continuum, experimental finding of material parameters is a harder problem than determining of elastic moduli in elasticity. This can be seen from the fact that an linear elastic isotropic material is determined by two elastic moduli whereas the description of an isotropic linear elastic micropolar material, Eqs. (4.67) and (4.68), needs six elastic parameters. We will touch the question of construction of the constitutive equations for a Cosserat continuum in brief.

1. Experimental data. Elastic moduli for micropolar materials can be determined from the experiments with specimens on tension, volume compression, torsion, flexure, and some dynamic experiments. For small deformations of homogeneous materials like steels, we cannot expect micropolar properties. They are watched in experiments with such micro-nonhomogeneous materials as foams, bones, rocks, granular media, etc. Besides micropolar properties arise at a neighborhood of stress concentrators like crack tips or notches. It seems that micropolar properties should be taken into account in nanoscale effects [71]. It should be recognized that the experimental data on micropolar solids are rather scarce. We can note the works on foams by Lakes [72–74] and on bones by Lakes et al. [75, 76]. In these, to define elastic micropolar modulus, it is used a so-called *size-effect* that is typical for a Cosserat continuum. For example, the size-effect becomes apparent in torsion tests on finding the shear modulus with specimens of few sizes. Lakes established the values of elastic modulus for some foams. Table 4.1 presents the modules for polystyrene (PS), dense polyurethane (PU), syntactic (Syn) foams, and for wet human bone.

We also should note experimental works by Gauthier and Jahsman [77, 78], who studied a composite material with aluminum shot uniformly distributed throughout an

Table 4.1 Micropolar moduli by Lakes [73, 74]

	Definition	Foam, PU	Foam, PS	Foam, Syn	Bone
Shear modulus (MPa)	$G = \dfrac{2\mu + \kappa}{2}$	1.1	104	1033	4000
Poisson's ratio	$\nu = \dfrac{\lambda}{2\lambda + 2\mu + \kappa}$	0.07	0.4	0.34	(no data)
Characteristic length (torsion) (mm)	$l_t = \sqrt{\dfrac{\beta_2 + \beta_3}{2\mu + \kappa}}$	3.8	0.62	0.065	0.22
Characteristic length (bending) (mm)	$l_b = \sqrt{\dfrac{\beta_3}{2(2\mu + \kappa)}}$	5.0	0.33	0.032	0.45
Coupling number	$N^2 = \dfrac{\kappa}{2\mu + \kappa}$	0.09	0.04	0.1	≥ 0.5
Polar ratio	$\Psi = \dfrac{\beta_2 + \beta_3}{\beta_1 + \beta_2 + \beta_3}$	1.5	1.5	1.5	1.5

epoxy matrix. Cosserat constants also are determined for polycarbonate honeycombs in [79]. The values of micropolar moduli can be found in [36, 80].

Using the concept of bounded stiffness and infinitesimal conformal invariance, Neff and Jeong [81, 82] and Neff et al. [83] presented the mathematical analysis of Cosserat constitutive equations that reduces the number of independent elastic moduli of a micropolar medium.

2. Homogenization. Various homogenization methods applied to lattices, granular medias, soils, cellular structures, etc. are a more significant source of known micropolar modules. The literature on this is more broad, we note the papers by Besdo [84], Bigoni and Drugan [85], Diebels [86], Dos Reis and Ganghoffer [87], Ehlers et al. [88], Forest et al. [89–91], Suiker et al. [92, 93], and [94–97]. The idea of homogenization is to replace a composite material or assembly of particles by an effective generalized continuum model. For example, consider a system of interacting rigid bodies for which the force and couple interaction are essential. Changing the system to an effective continuum medium we expect the medium to inherit couple properties and rotational freedom degrees of rigid bodies. Changing open-cell foam that can be considered as a system of elastic beams to an effective continuum medium, we get a similar modeling situation. Thus the homogenization technique for a micro-inhomogeneous medium replaces direct experiments and can subject us with the values of micropolar modules.

Note that the homogenization can give us more general models of continuum medium like micromorphic continuum, see, for example, the papers by Forest and Sievert [91], Forest and Duy [98], Kouznetsova et al. [99], Neff and Forest [100], and Cielecka et al. [101].

3. Numerics. At first glance, numerical methods are an unexpected source of constitutive equations for the Cosserat continuum. Here Cosserat constitutive equations are used to accelerate numerical solving of classical problem of elasticity or elasto-plasticity. The choice of the elastic moduli is determined by the need to accelerate convergence to a solution or to regularize the solution. Note that the introduction of the rotations as independent kinematic variables for more accurate description of 3D continuum relates to Reissner's idea in the plate theory. In this approach, micropolar modules are a kind of numerical tuning values, they have no certain physical meaning. In elasticity and elasto-plasticity, the approach is used by Neff et al. [102, 103].

Among other numerical works that used the idea of the Cosserat continuum we should note [104–108].There were developed special types of enhanced finite elements that use the representation of a finite element as *a Cosserat point*.

4. Micropolar fluids. For micropolar fluids, the situation of constitutive equations is different. Starting with pioneering papers by Aero et al. [63] and Eringen [64] the micropolar continuum is applied to model magnetic liquids, polymer suspensions, liquid crystals, and other types of fluids with microstructure, see also the books by Migoun and Prokhorenko [65], Łukaszewicz [66], Eringen [67], and Eremeyev and Zubov [109]. In particular, the use of magnetic fluids named also ferrofluids [110] caused the development of micropolar hydromechanics where a magnetic field induces voluminous couples. To describe the couples, liquid corpuscules should possess rotational degrees of freedom. Comparing with micropolar elasticity, micropolar hydrodynamics is a more extensive part of mechanics with well established experimentally constitutive equations.

References

1. C. Truesdell, W. Noll, The nonlinear field theories of mechanics. in *Handbuch der Physik*, vol. III/3, ed. by S. Flügge (Springer, Berlin, 1965), pp. 1–602
2. A.C. Eringen, *Nonlocal Continuum Field Theories* (Springer, New York, 2002)
3. C.B. Kafadar, A.C. Eringen, Micropolar media—I. The classical theory. Int. J. Eng. Sci. **9**(3), 271–305 (1971)
4. K.C. Le, H. Stumpf, Strain measures, integrability condition and frame indifference in the theory of oriented media. Int. J. Solids Struct. **35**(9–10), 783–798 (1998)
5. W. Noll, A mathematical theory of the mechanical behavior of continuous media. Arch. Ration. Mech. Anal. **2**(1), 197–226 (1958)
6. A.I. Murdoch, On objectivity and material symmetry for simple elastic solids. J. Elast. **60**(3), 233–242 (2000)
7. A.I. Murdoch, Objectivity in classical continuum physics: a rationale for discarding the 'principle of invariance under superposed rigid body motions' in favour of purely objective considerations. Continuum Mech. Thermodyn. **15**(3), 309–320 (2003)
8. A.I. Murdoch, On criticism of the nature of objectivity in classical continuum physics. Continuum Mech. Thermodyn. **17**(2), 135–148 (2005)
9. W. Muschik, L. Restuccia, Changing the observer and moving materials in continuum physics: objectivity and frame-indifference. Technische Mechanik **22**(2), 152–160 (2002)
10. R.S. Rivlin, Material symmetry revisited. GAMM Mitteilungen **25**(1/2), 109–126 (2002)

11. R.S. Rivlin, Frame indifference and relative frame indifference. Math. Mech. Solids **10**(2), 145–154 (2005)
12. R.S. Rivlin, Some thoughts on frame indifference. Math. Mech. Solids **11**(2), 113–122 (2006)
13. A. Bertram, B. Svendsen, On material objectivity and reduced constitutive equations. Arch. Mech. **53**(6), 653–675 (2001)
14. A. Bertram, B. Svendsen, Reply to Rivlin's material symmetry revisited or much ado about nothing. GAMM Mitteilungen **27**(1), 88–93 (2004)
15. B. Svendsen, A. Bertram, On frame-indifference and form-invariance in constitutive theory. Acta Mechanica **132**(1–4), 195–207 (1999)
16. A. Bertram, *Elasticity and Plasticity of Large Deformations: An Introduction*, 2nd edn. (Springer, Berlin, 2008)
17. R.P. Feynman, R.B. Leighton, M. Sands, *The Feynman Lectures on Physics*, vol. 1, 6th edn. (Addison-Wesley, Reading, 1977)
18. W. Pietraszkiewicz, V.A. Eremeyev, On natural strain measures of the non-linear micropolar continuum. Int. J. Solids Struct. **46**(3–4), 774–787 (2009)
19. W. Pietraszkiewicz, V.A. Eremeyev, On vectorially parameterized natural strain measures of the non-linear Cosserat continuum. Int. J. Solids Struct. **46**(11–12), 2477–2480 (2009)
20. E. Cosserat, F. Cosserat, *Théorie des corps déformables* (Herman et Fils, Paris, 1909)
21. T. Merlini, A variational formulation for finite elasticity with independent rotation and biot-axial fields. Comput. Mech. **19**(3), 153–168 (1997)
22. L.M. Zubov, V.A. Eremeev, Equations for a viscoelastic micropolar fluid. Doklady Phys. **41**(12), 598–601 (1996)
23. V.A. Yeremeyev, L.M. Zubov, The theory of elastic and viscoelastic micropolar liquids. J. Appl. Math. Mech. **63**(5), 755–767 (1999)
24. J. Chróscielewski, J. Makowski, W. Pietraszkiewicz, Statics and dynamics of multyfolded shells. in *Nonlinear Theory and Finite Elelement Method (in Polish)* (Wydawnictwo IPPT PAN, Warszawa, 2004)
25. V.A. Eremeyev, W. Pietraszkiewicz, Local symmetry group in the general theory of elastic shells. J. Elast. **85**(2), 125–152 (2006)
26. V.A. Eremeyev, W. Pietraszkiewicz, Material symmetry group of the non-linear polar-elastic continuum. Int. J. Solids Struct. **49**(14), 1993–2005 (2012)
27. E. Reissner, On kinematics and statics of finite-strain force and moment stress elasticity. Stud. Appl. Math. **52**, 93–101 (1973)
28. E. Reissner, Note on the equations of finite-strain force and moment stress elasticity. Stud. Appl. Math. **54**, 1–8 (1975)
29. R. Stojanovic, Nonlinear micropolar elasticity, in *Micropolar Elasticity*, vol. 151, ed. by W. Nowacki, W. Olszak (Springer, Wien, 1974), pp. 73–103
30. L.M. Zubov, Variational principles and invariant integrals for non-linearly elastic bodies with couple stresses (in Russian). Mech. Solids (6), 10–16 (1990)
31. V.A. Eremeyev, L.M. Zubov, On the stability of elastic bodies with couple stresses. Mech. Solids **29**(3), 172–181 (1994)
32. L.M. Zubov, *Nonlinear Theory of Dislocations and Disclinations in Elastic Bodies* (Springer, Berlin, 1997)
33. E. Nikitin, L.M. Zubov, Conservation laws and conjugate solutions in the elasticity of simple materials and materials with couple stress. J. Elast. **51**(1), 1–22 (1998)
34. O.A. Bauchau, L. Trainelli, The vectorial parameterization of rotation. Nonlinear Dyn. **32**(1), 71–92 (2003)
35. J. Dyszlewicz, *Micropolar Theory of Elasticity* (Springer, Berlin, 2004)
36. A.C. Eringen, *Microcontinuum Field Theory I. Foundations and Solids* (Springer, New York, 1999)
37. W. Nowacki, *Theory of Asymmetric Elasticity* (Pergamon-Press, Oxford, 1986)
38. C. Truesdell, Die Entwicklung des Drallsatzes. ZAMM **44**(4/5), 149–158 (1964)
39. C. Truesdell, *A First Course in Rational Continuum Mechanics* (Academic Press, New York, 1977)

40. C.C. Wang, C. Truesdell, *Introduction to Rational Elasticity* (Noordhoof Int. Publishing, Leyden, 1973)
41. R.S. Rivlin, Material symmetry and constitutive equations. Ingenieur-Archiv **49**(5–6), 325–336 (1980)
42. A.C. Eringen, C.B. Kafadar, Polar field theories. in *Continuum Physics*, vol. IV, ed. by A.C. Eringen, (Academic Press, New York 1976), pp. 1–75
43. A.J.M. Spencer, Isotropic integrity bases for vectors and second-order tensors. Part II. Arch. Ration. Mech. Anal. **18**(1), 51–82 (1965)
44. A.J.M. Spencer, Theory of invariants, in *Continuum Physics*, vol. 1, ed. by A.C. Eringen (Academic Press, New-York, 1971), pp. 239–353
45. Q.S. Zheng, Theory of representations for tensor functions—a unified invariant approach to constitutive equations. Appl. Mech. Rev. **47**(11), 545–587 (1994)
46. S. Ramezani, R. Naghdabadi, S. Sohrabpour, Constitutive equations for micropolar hyper-elastic materials. Int. J. Solids Struct. **46**(14–15), 2765–2773 (2009)
47. Q.S. Zheng, A.J.M. Spencer, On the canonical representations for Kronecker powers of orthogonal tensors with application to material symmetry problems. Int. J. Eng. Sci. **31**(4), 617–635 (1993)
48. H. Xiao, On symmetries and anisotropies of classical and micropolar linear elasticities: a new method based upon a complex vector basis and some systematic results. J. Elast. **49**(2), 129–162 (1998)
49. A.I. Lurie, *Theory of Elasticity* (Springer, Berlin, 2005)
50. R.W. Ogden, *Non-linear Elastic Deformations* (Ellis Horwood, Chichester, 1984)
51. A.I. Lurie, *Nonlinear Theory of Elasticity* (North-Holland, Amsterdam, 1990)
52. A.O. Varga, *Stress-Strain Behavior of Elastic Materials: Selected Problems of Large Deformations* (Interscience, New York, 1996)
53. C. Truesdell, *Rational Thermodynamics*, 2nd edn. (Springer, New York, 1984)
54. J.K. Knowles, E. Sternberg, On the ellipticity of the equation of nonlinear elastostatics for a special material. J. Elast. **5**(3–4), 341–361 (1975)
55. J.K. Knowles, E. Sternberg, On the failure of ellipticity and the emergence of discontinuous deformation gradients in plane finite elastostatics. J. Elast. **10**, 255–293 (1980)
56. P. Rosakis, Ellipticity and deformation with discontinuous gradients in finite elastostatics. Arch. Ration. Mech. Anal. **109**(1), 1–37 (1990)
57. L. Zee, E. Sternberg, Ordinary and strong ellipticity in the equilibrium theory of incompressible hyperelastic solids. Arch. Ration. Mech. Anal. **83**(1), 53–90 (1983)
58. J.L. Lions, E. Magenes, *Problèmes aux limites non homogènes et applications* (Dunod, Paris, 1968)
59. G. Fichera, Existence theorems in elasticity. in *Handbuch der Physik*, vol. VIa/2, ed. by S. Flügge, (Springer, Berlin 1972), pp. 347–389
60. L. Hörmander, *Linear Partial Differential Equations, A Series of Comprehensive Studies in Mathematics*, vol. 116, 4th edn. (Springer, Berlin, 1976)
61. M. Agranovich, Elliptic boundary problems. in *Partial Differential Equations IX: Elliptic Boundary Problems. Encyclopaedia of Mathematical Sciences*, vol. 79, ed. by M. Agranovich, Y. Egorov, M. Shubin (Springer, Berlin, 1997), pp. 1–144
62. L. Nirenberg, *Topics in Nonlinear Functional Analysis* (American Mathematical Society, New York, 2001)
63. E.L. Aero, A.N. Bulygin, E.V. Kuvshinskii, Asymmetric hydromechanics. J. Appl. Math. Mech. **29**(2), 333–346 (1965)
64. A.C. Eringen, Theory of micropolar fluids. J. Math. Mech. **16**(1), 1–18 (1966)
65. N.P. Migoun, P.P. Prokhorenko, *Hydrodynamics and Heattransfer in Gradient Flows of Microstructured Fluids (in Russian)* (Nauka i Technika, Minsk, 1984)
66. G. Łukaszewicz, *Micropolar Fluids: Theory and Applications* (Birkhäuser, Boston, 1999)
67. A.C. Eringen, *Microcontinuum Field Theory. II. Fluent Media* (Springer, New York, 2001)
68. A.C. Eringen, A unified continuum theory of liquid crystals. ARI Int. J. Phys. Eng. Sci. **73–84**(2), 369–374 (1997)

69. A.C. Eringen, A unified continuum theory of electrodynamics of liquid crystals. Int. J. Eng. Sci. **35**(12–13), 1137–1157 (1997)
70. A.C. Eringen, A unified continuum theory for electrodynamics of polymeric liquid crystals. Int. J. Eng. Sci. **38**(9–10), 959–987 (2000)
71. E.A. Ivanova, A.M. Krivtsov, N.F. Morozov, A.D. Firsova, Description of crystal packing of particles with torque interaction. Mech. Solids **38**(4), 76–88 (2003)
72. R.S. Lakes, Experimental microelasticity of two porous solids. Int. J. Solids Struct. **22**(1), 55–63 (1986)
73. R.S. Lakes, Experimental micro mechanics methods for conventional and negative Poisson's ratio cellular solids as Cosserat continua. Trans. ASME J. Eng. Mater. Technol. **113**(1), 148–155 (1991)
74. R.S. Lakes, Experimental methods for study of Cosserat elastic solids and other generalized continua. in *Continuum Models for Materials with Micro-Structure*, ed. by H. Mühlhaus (Wiley, New York, 1995), pp. 1–22
75. H.C. Park, R.S. Lakes, Cosserat micromechanics of human bone: strain redistribution by a hydration-sensitive constituent. J. Biomech. **19**(5), 385–397 (1986)
76. J.F.C. Yang, R.S. Lakes, Experimental study of micropolar and couple stress elasticity in compact bone in bending. J. Biomech. **15**(2), 91–98 (1982)
77. R.D. Gauthier, W.E. Jahsman, Quest for micropolar elastic-constants. Trans. ASME J. Appl. Mech. **42**(2), 369–374 (1975)
78. R.D. Gauthier, W.E. Jahsman, Quest for micropolar elastic-constants. 2. Arch. Mech. **33**(5), 717–737 (1981)
79. R. Mora, A.M. Waas, Measurement of the Cosserat constant of circular-cell polycarbonate honeycomb. Philos. Mag. A. Phys. Condens. Matter Struct. Defects Mech. Prop. **80**(7), 1699–1713 (2000)
80. V.I. Erofeev, *Wave Processes in Solids with Microstructure* (World Scientific, Singapore, 2003)
81. P. Neff, J. Jeong, A new paradigm: the linear isotropic Cosserat model with conformally invariant curvature energy. ZAMM **89**(2), 107–122 (2009)
82. J. Jeong, P. Neff, Existence, uniqueness and stability in linear Cosserat elasticity for weakest curvature conditions. Math. Mech. Solids **15**(1), 78–95 (2010)
83. P. Neff, J. Jeong, A. Fischle, Stable identification of linear isotropic Cosserat parameters: bounded stiffness in bending and torsion implies conformal invariance of curvature. Acta Mechanica **211**(3–4), 237–249 (2010)
84. D. Besdo, Towards a Cosserat-theory describing motion of an originally rectangular structure of blocks. Arch. Appl. Mech. **80**(1), 25–45 (2010)
85. D. Bigoni, W.J. Drugan, Analytical derivation of Cosserat moduli via homogenization of heterogeneous elastic materials. Trans. ASME J. Appl. Mech. **74**(4), 741–753 (2007)
86. S. Diebels, A micropolar theory of porous media: constitutive modelling. Transp. Porous Media **34**(1–3), 193–208 (1999)
87. F. Dos Reis, J.F. Ganghoffer, Construction of micropolar continua from the homogenization of repetitive planar lattices. in *Mechanics of Generalized Continua, Advanced Structured Materials*, vol. 7, ed. by H. Altenbach, G.A. Maugin, V. Erofeev (Springer, Berlin, 2011), pp. 193–217
88. W. Ehlers, E. Ramm, S. Diebels, G.D.A. d'Addetta, From particle ensembles to Cosserat continua: homogenization of contact forces towards stresses and couple stresses. Int. J. Solids Struct. **40**(24), 6681–6702 (2003)
89. S. Forest, Mechanics of generalized continua: construction by homogenizaton. J. de Physique IV France 8(PR4), Pr4-39–Pr4-48 (1998)
90. S. Forest, K. Sab, Cosserat overal modelling of heterogeneous materials. Mech. Res. Commun. **25**(4), 449–454 (1998)
91. S. Forest, R. Sievert, Nonlinear microstrain theories. Int. J. Solids Struct. **43**(24), 7224–7245 (2006)

92. A.S.J. Suiker, A.V. Metrikine, R. de Borst, Comparison of wave propagation characteristics of the Cosserat continuum model and corresponding discrete lattice models. Int. J. Solids Struct. **38**(9), 1563–1583 (2001)
93. A.S.J. Suiker, R. de Borst, Enhanced continua and discrete lattices for modelling granular assemblies. Philos. Trans. Royal Soc. A **363**(1836), 2543–2580 (2005)
94. R. Larsson, S. Diebels, A second-order homogenization procedure for multi-scale analysis based on micropolar kinematics. Int. J. Numer. Methods Eng. **69**(12), 2485–2512 (2007)
95. R. Larsson, Y. Zhang, Homogenization of microsystem interconnects based on micropolar theory and discontinuous kinematics. J. Mech. Phys. Solids **55**(4), 819–841 (2007)
96. E. Pasternak, H.B. Mühlhaus, Generalised homogenisation procedures for granular materials. J. Eng. Math. **52**(1), 199–229 (2005)
97. O. van der Sluis, P.H.J. Vosbeek, P.J.G. Schreurs, H.E.H. Meijer, Homogenization of heterogeneous polymers. Int. J. Solids Struct. **36**(21), 3193–3214 (1999)
98. S. Forest, K.T. Duy, Generalized continua and non-homogeneous boundary conditions in homogenisation methods. ZAMM **91**(2), 90–109 (2011)
99. V. Kouznetsova, M.G.D. Geers, W.A.M. Brekelmans, Multi-scale constitutive modelling of heterogeneous materials with a gradient-enhanced computational homogenization scheme. Int. J. Numer. Methods Eng. **54**(8), 1235–1260 (2002)
100. P. Neff, S. Forest, A geometrically exact micromorphic model for elastic metallic foams accounting for affine microstructure. Modelling, existence of minimizers, identification of moduli and computational results. J. Elast. **87**(2–3), 239–276 (2007)
101. I. Cielecka, M. Wozniak, C. Wozniak, Elastodynamic behaviour of honeycomb cellular media. J. Elast. **60**(1), 1–17 (2000)
102. P. Neff, K. Chełminski, W. Müller, C. Wieners, A numerical solution method for an infinitesimal elasto-plastic Cosserat model. Math. Models Methods Appl. Sci. **17**(8), 1211–1239 (2007)
103. P. Neff, K. Chełminski, Well-posedness of dynamic Cosserat plasticity. Appl. Math. Optim. **56**(1), 19–35 (2007)
104. M.B. Rubin, On the theory of a Cosserat point and its application to the numerical solution of continuum problems. Trans. ASME J. Appl. Mech. **52**(2), 368–372 (1985)
105. M.B. Rubin, *Cosserat Theories: Shells Rods and Points* (Kluwer, Dordrecht, 2000)
106. B. Nadler, M. Rubin, A new 3-D finite element for nonlinear elasticity using the theory of a Cosserat point. Int. J. Solids Struct. **40**(17), 4585–4614 (2003)
107. M. Jabareen, M. Rubin, An improved 3-D brick Cosserat point element for irregular shaped elements. Comput. Mech. **40**(6), 979–1004 (2007)
108. M. Jabareen, M. Rubin, Modified torsion coefficients for a 3-D brick Cosserat point element. Comput. Mech. **41**(4), 517–525 (2008)
109. V.A. Eremeyev, L.M. Zubov, *Principles of Viscoelastic Micropolar Fluid Mechanics (in Russian)* (SSC of RASci Publishers, Rostov on Don, 2009)
110. R.E. Rosensweig, Magnetic fluids. Ann. Rev. Fluid Mech. **19**, 437–461 (1987)

Chapter 5
Strong Ellipticity and Acceleration Waves in Micropolar Continuum

In this chapter we will consider acceleration waves in nonlinear thermoelastic micropolar continua. We will establish kinematic and dynamic compatibility relations for a singular surface of second order in the media. We also will derive an analogue to the Fresnel–Hadamard–Duhem theorem and an expression for the acoustic tensor. The condition for acceleration wave's propagation is formulated as an algebraic spectral problem. It is shown that the condition coincides with the strong ellipticity of equilibrium equations. As an example, a quadratic form for the specific free energy is considered and solutions to the corresponding spectral problem are presented.

The propagation of nonlinear waves in solids is a complex process and analytic solutions to nonlinear problems are rare. However, the problem of propagation of acceleration waves is exceptional and we can present some analytic results. An *acceleration wave* (or a *wave of weak discontinuity of order* 2) is a solution to the motion equations of continuum that possesses discontinuities in the second derivatives on some surfaces that will be called *singular*. It means that the acceleration wave is represented by a traveling surface which is a carrier of discontinuity jumps of the second derivatives of a solution with respect to the spacial coordinates and time whereas the solution and its first derivatives are continuous some surface neighborhood. From the mathematical point of view, existence of acceleration waves closely relates with *hyperbolicity* of the dynamic equations or their ellipticity for the equilibrium equations. Existence of acceleration waves in any direction is equivalent to the fact that all eigenvalues of an algebraic spectral problem for the acoustic tensor are positive for any direction of wave propagation.

From the physical point of view, the hyperbolicity of the equations of motion is a natural property of dynamics of elastic media as well as ellipticity is a natural property of its statics. The violation of hyperbolicity (or elipticity) means that discontinuous solutions may appear. Such solutions may model shear-bands, phase transitions, interfaces, fracture, defects, slip surfaces and other phenomena. So, the algebraic criterion for such phenomena is important in mechanics of materials.

V. A. Eremeyev et al., *Foundations of Micropolar Mechanics*,
SpringerBriefs in Continuum Mechanics,
DOI: 10.1007/978-3-642-28353-6_5,

The investigations of acceleration waves in nonlinear elastic and thermoelastic media were performed in many works, see [1–6]. Acceleration waves in elastic micropolar media were considered in [7]. A generalization for elastic and viscoelastic micropolar medias was presented in [8]. The relation between the existence of acceleration waves and the condition of strong ellipticity of the equilibrium equations was established in [9]. Formulation of the strong ellipticity condition for the micropolar media was given in [10], see Sect. 4.9. The derivation of acoustic tensor for micropolar media with application to localization phenomena in micropolar elastoplasticity was done by [11]. We should also mention the monographs [12–15], where the wave processes in micropolar continua are considered.

5.1 Thermoconductivity Equation in the Micropolar Continuum

When there is heat transfer in a media, its motion equations (3.39) should be supplemented with the Lagrangian heat conduction equation [12, 16]

$$\rho_0 \theta \frac{\mathrm{d}\eta}{\mathrm{d}t} = -\mathrm{Div}\,\mathbf{q} + \rho h, \tag{5.1}$$

where θ is the temperature, η the specific entropy, $\mathbf{q}$ the heat flux in the reference configuration, and h the density of external heat source.

The constitutive equations for a Cosserat thermoelastic continuum can be derived with the use of the specific free energy

$$\psi = \psi(\mathbf{E}, \mathbf{K}, \theta) \tag{5.2}$$

as follows

$$\mathbf{T}_\kappa = \rho_0 \mathbf{H} \cdot \psi_{,\mathbf{E}}, \quad \mathbf{M}_\kappa = \rho_0 \mathbf{H} \cdot \psi_{,\mathbf{K}}, \quad \eta = -\psi_{,\theta}, \quad \mathbf{q} = \mathbf{q}(\mathbf{E}, \mathbf{K}, \theta, \mathrm{Grad}\,\theta). \tag{5.3}$$

As is known, from the second law of thermodynamics it follows $\mathbf{q} \cdot \mathrm{Grad}\,\theta \leq 0$.

From now on, we assume ψ to be a twice continuously differentiable function and vector-function $\mathbf{q}$ to be continuously differentiable.

5.2 Acceleration Waves

We consider motions of the continuum when discontinuities of kinematic and dynamic quantities appear at a smooth surface $S(t)$ that is called singular (Fig. 5.1). For the quantities describing the motion at $S(t)$, we suppose existence of unilateral limit values. For the second derivatives of the motion, the limits from each of both sides from $S(t)$ can differ, in general. A jump for a quantity at $S(t)$ is denoted by the double square brackets, for example, $[\![\psi]\!] = \psi^+ - \psi^-$.

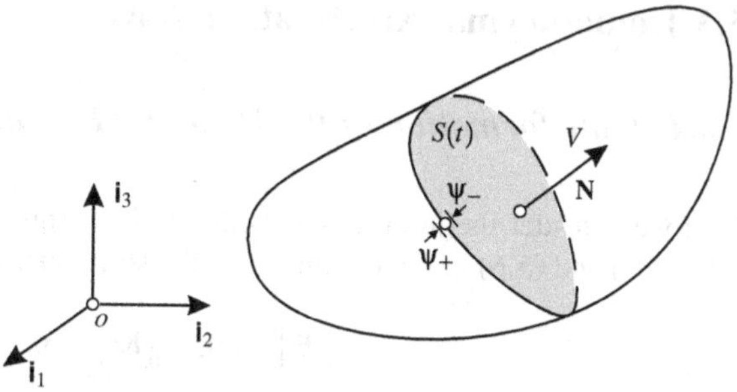

Fig. 5.1 Singular surface

Under these assumptions, let us derive the conditions for existence of the acceleration wave, that is the conditions under which weak discontinuous solutions can arise. On the singular surface the following balance equations must be valid (see [12, 16]):

$$\rho_0 V [\![\mathbf{v}]\!] = -[\![\mathbf{T}_\kappa]\!] \cdot \mathbf{N}, \quad \rho_0 \gamma V [\![\boldsymbol{\omega}]\!] = -[\![\mathbf{M}_\kappa]\!] \cdot \mathbf{N}, \quad \rho_0 \theta V [\![\eta]\!] = [\![\mathbf{q}]\!] \cdot \mathbf{N}, \tag{5.4}$$

where V is the intrinsic speed of propagation of $S(t)$ in the direction $\mathbf{N}$ [2–4] and $\mathbf{N}$ is the unit normal to S.

Definition 5.1 An acceleration wave (*or weak discontinuity wave, or singular surface of the second order*) is a traveling singular surface $S(t)$ at which the second spatial and time derivatives of the position vector $\mathbf{r}$ and of the micro-rotation tensor $\mathbf{H}$ have jumps, while $\mathbf{r}$ and $\mathbf{H}$ together with all their first derivatives are continuous.

So on $S(t)$ we have

$$[\![\mathbf{F}]\!] = \mathbf{0}, \quad [\![\operatorname{Grad} \mathbf{H}]\!] = \mathbf{0}, \quad [\![\mathbf{v}]\!] = \mathbf{0}, \quad [\![\boldsymbol{\omega}]\!] = \mathbf{0}. \tag{5.5}$$

From (5.4) and (5.5) it follows $[\![\mathbf{T}_\kappa]\!] \cdot \mathbf{N} = \mathbf{0}$, $[\![\mathbf{M}_\kappa]\!] \cdot \mathbf{N} = \mathbf{0}$.

Let us consider two types of acceleration waves. The first one is the *homothermal* acceleration wave when *the temperature field and its first derivatives are continuous at* $S(t)$:

$$[\![\theta]\!] = 0, \quad [\![\operatorname{Grad} \theta]\!] = \mathbf{0}, \quad [\![\dot{\theta}]\!] = 0. \tag{5.6}$$

The second type is the *homentropic* (or *homocaloric*) acceleration wave when *the entropy field and its first derivatives are continuous at* $S(t)$:

$$[\![\eta]\!] = 0, \quad [\![\operatorname{Grad} \eta]\!] = \mathbf{0}, \quad [\![\dot{\eta}]\!] = 0. \tag{5.7}$$

For the homentropic acceleration wave, the Fourier condition holds:

$$[\![\mathbf{q}]\!] \cdot \mathbf{N} = 0. \tag{5.8}$$

5.3 Homothermal Acceleration Waves

5.3.1 Transformation of the Dynamic Equations

First we consider the jump relations that follow from the motion equations. Equations (5.5) and (5.6) imply continuity of the strain measures $\mathbf{E}$ and $\mathbf{K}$ at $S(t)$

$$[\![\mathbf{E}]\!] = \mathbf{0}, \quad [\![\mathbf{K}]\!] = \mathbf{0}.$$

Hence, in view of the constitutive equations (5.2), (5.3), we establish the continuity of the tensors $\mathbf{T}_\kappa$ and $\mathbf{M}_\kappa$, of the entropy density, and of the heat flow vector:

$$[\![\mathbf{T}_\kappa]\!] = \mathbf{0}, \quad [\![\mathbf{M}_\kappa]\!] = \mathbf{0}, \quad [\![\eta]\!] = 0, \quad [\![\mathbf{q}]\!] = \mathbf{0}.$$

It follows immediately the balance equations (5.4) to be valid on $S(t)$.

Let us recall *Maxwell's theorem*, see [2–4]:

Theorem 5.1 (Maxwell) *For a continuously differentiable field* $\boldsymbol{\psi}$ *such that* $[\![\boldsymbol{\psi}]\!] = \mathbf{0}$ *the following relations hold*

$$[\![\dot{\boldsymbol{\psi}}]\!] = -V\boldsymbol{\varphi}, \quad [\![\mathrm{Grad}\,\boldsymbol{\psi}]\!] = \boldsymbol{\varphi} \otimes \mathbf{N}, \tag{5.9}$$

where $\boldsymbol{\varphi}$ *is the tensor amplitude of the jump of the first gradient of* $\boldsymbol{\psi}$*; the tensor amplitude is a tensor of the order equal to the order of* $\boldsymbol{\psi}$.

Application of Maxwell's theorem to the continuous fields of $\mathbf{v}$, $\boldsymbol{\omega}$, $\mathbf{T}_\kappa$, and $\mathbf{M}_\kappa$ yield in the system of equations that relate the jumps in the derivatives with respect to the spatial variables and time t at $S(t)$:

$$[\![\dot{\mathbf{v}}]\!] = -V\mathbf{a}, \quad [\![\mathrm{Grad}\,\mathbf{v}]\!] = \mathbf{a} \otimes \mathbf{N}, \quad [\![\dot{\boldsymbol{\omega}}]\!] = -V\mathbf{b}, \quad [\![\mathrm{Grad}\,\boldsymbol{\omega}]\!] = \mathbf{b} \otimes \mathbf{N},$$
$$V\,[\![\mathrm{Div}\,\mathbf{T}_\kappa]\!] = -\left[\!\!\left[\dot{\mathbf{T}}_\kappa\right]\!\!\right] \cdot \mathbf{N}, \quad V\,[\![\mathrm{Div}\,\mathbf{M}_\kappa]\!] = -\left[\!\!\left[\dot{\mathbf{M}}_\kappa\right]\!\!\right] \cdot \mathbf{N},$$

where $\mathbf{a}$ and $\mathbf{b}$ are the vectorial amplitudes of the jumps in the linear and angular accelerations. With this, we get the relations for the jumps in the derivatives of the strain measures with respect to t

$$\left[\!\!\left[\dot{\mathbf{E}}\right]\!\!\right] = \mathbf{H}^T \cdot \mathbf{a} \otimes \mathbf{N}, \quad \left[\!\!\left[\dot{\mathbf{K}}\right]\!\!\right] = \mathbf{H}^T \cdot \mathbf{b} \otimes \mathbf{N}. \tag{5.10}$$

Let the mass forces and couples be continuous. From the balance equations (3.39) it follows

$$[\![\mathrm{Div}\,\mathbf{T}_\kappa]\!] = \rho_0\,[\![\dot{\mathbf{v}}]\!], \quad [\![\mathrm{Div}\,\mathbf{M}_\kappa]\!] = \rho_0\gamma\,[\![\dot{\boldsymbol{\omega}}]\!]. \tag{5.11}$$

Using Eqs. (5.10) and (5.11) we get the relations

$$-\left[\!\left[\dot{\mathbf{T}}_\kappa\right]\!\right]\cdot\mathbf{N}=\rho_0 V\,[\dot{\mathbf{v}}]\,,\quad -\left[\!\left[\dot{\mathbf{M}}_\kappa\right]\!\right]\cdot\mathbf{N}=\rho_0\gamma V\,[\![\dot{\boldsymbol{\omega}}]\!]\,. \tag{5.12}$$

Differentiating the constitutive equations (5.3) we obtain

$$\left[\!\left[\dot{\mathbf{T}}_\kappa\right]\!\right]=\rho_0\psi_{,\mathbf{EE}}\cdot\cdot\left[\!\left[\dot{\mathbf{E}}\right]\!\right]^T+\rho_0\psi_{,\mathbf{EK}}\cdot\cdot\left[\!\left[\dot{\mathbf{K}}\right]\!\right]^T,$$
$$\left[\!\left[\dot{\mathbf{M}}_\kappa\right]\!\right]=\rho_0\psi_{,\mathbf{KE}}\cdot\cdot\left[\!\left[\dot{\mathbf{E}}\right]\!\right]^T+\rho_0\psi_{,\mathbf{KK}}\cdot\cdot\left[\!\left[\dot{\mathbf{K}}\right]\!\right]^T.$$

Next, with regard for (5.10), we transform (5.12) into the form containing the vectorial amplitudes **a** and **b** only:

$$\left(\psi_{,\mathbf{EE}}\cdot\cdot\left(\mathbf{N}\otimes\mathbf{H}^T\cdot\mathbf{a}\right)\right)\cdot\mathbf{N}+\left(\psi_{,\mathbf{EK}}\cdot\cdot\left(\mathbf{N}\otimes\mathbf{H}^T\cdot\mathbf{b}\right)\right)\cdot\mathbf{N}=V^2\mathbf{H}^T\cdot\mathbf{a},$$
$$\left(\psi_{,\mathbf{KE}}\cdot\cdot\left(\mathbf{N}\otimes\mathbf{H}^T\cdot\mathbf{a}\right)\right)\cdot\mathbf{N}+\left(\psi_{,\mathbf{KK}}\cdot\cdot\left(\mathbf{N}\otimes\mathbf{H}^T\cdot\mathbf{b}\right)\right)\cdot\mathbf{N}=\gamma V^2\mathbf{H}^T\cdot\mathbf{b}.$$

Using matrix notation, we rewrite these in a more compact form:

$$\mathbb{Q}(\mathbf{N})\cdot\boldsymbol{\xi}=V^2\mathbb{B}\cdot\boldsymbol{\xi}, \tag{5.13}$$

where

$$\boldsymbol{\xi}=(\mathbf{a}',\mathbf{b}')\in\mathbb{R}^6,\quad \mathbf{a}'=\mathbf{H}^T\cdot\mathbf{a},\quad \mathbf{b}'=\mathbf{H}^T\cdot\mathbf{b},$$

$$\mathbb{Q}(\mathbf{N})\equiv\begin{bmatrix}\psi_{,\mathbf{EE}}\{\mathbf{N}\} & \psi_{,\mathbf{EK}}\{\mathbf{N}\}\\ \psi_{,\mathbf{KE}}\{\mathbf{N}\} & \psi_{,\mathbf{KK}}\{\mathbf{N}\}\end{bmatrix},\quad \mathbb{B}\equiv\begin{bmatrix}\mathbf{I} & \mathbf{0}\\ \mathbf{0} & \gamma\mathbf{I}\end{bmatrix}.$$

$\mathbb{Q}(\mathbf{N})$ is an analogue to the *homothermal acoustic tensor* for the micropolar continuum. From existence of the free energy function ψ it follows that $\mathbb{Q}(\mathbf{N})$ is symmetric. This provides that the squared velocity of propagation for an acceleration wave in an elastic micropolar continuum is real-valued. The requirement that $\mathbb{Q}(\mathbf{N})$ has to be positive definite is necessary for existence of an acceleration wave. It coincides with the condition of strong ellipticity of the equilibrium equations for an elastic micropolar continuum [9].

5.3.2 Transformation of the Heat Conductivity Equation

Now let us consider how the temperature field affects the existence of acceleration waves: we derive some relations for the jumps. Applying Maxwell's theorem to the field of **q** and to the temperature gradient, we get

$$V\,[\![\mathrm{Div}\,\mathbf{q}]\!] = -[\![\dot{\mathbf{q}}]\!]\cdot\mathbf{N},\quad [\![\mathrm{Grad}\,\mathrm{Grad}\,\theta]\!] = \mathbf{g}\otimes\mathbf{N},\quad \left[\!\left[\dot{(\mathrm{Grad}\,\theta)}\right]\!\right] = -V\mathbf{g}, \tag{5.14}$$

where $\mathbf{g}$ is the vector amplitude of the jump in the second gradient of the temperature. Similar to (5.10), from (5.1) it follows that

$$[\![\mathrm{Div}\,\mathbf{q}]\!] = -\rho_0\theta\,[\![\dot{\eta}]\!]\,. \tag{5.15}$$

From (5.14) and (5.15), we obtain

$$[\![\dot{\mathbf{q}}]\!]\cdot\mathbf{N} = \rho_0\theta\,[\![\dot{\eta}]\!]\,. \tag{5.16}$$

Let us restrict further discussion by the assumption that the constitutive equation for $\mathbf{q}$ obeys *Fourier's law*

$$\mathbf{q} = -\mathbf{k}(\theta)\cdot\mathrm{Grad}\,\theta,\quad \mathbf{h}\cdot\mathbf{k}(\theta)\cdot\mathbf{h} > 0,\quad \forall\mathbf{h}\neq\mathbf{0}, \tag{5.17}$$

where $\mathbf{k}$ is the positive definite thermoconductivity tensor. Differentiating $(5.3)_3$ and (5.17) with respect to t and using (5.16), we get

$$\mathbf{N}\cdot\mathbf{k}(\theta)\cdot\mathbf{g} = \rho_0\theta\left(\mathbf{a}'\cdot\psi_{,\theta\mathbf{E}}\cdot\mathbf{N} + \mathbf{b}'\cdot\psi_{,\theta\mathbf{K}}\cdot\mathbf{N}\right). \tag{5.18}$$

Now, using again the matrix notation, we can rewrite (5.13) and (5.18)

$$\mathbb{Q}_\theta(\mathbf{N})\cdot\boldsymbol{\zeta} = V^2\mathbb{B}_\theta\cdot\boldsymbol{\zeta}, \tag{5.19}$$

where $\boldsymbol{\zeta} = (\mathbf{a}', \mathbf{b}', \mathbf{g}) \in \mathbb{R}^9$, $\mathbb{Q}_\theta$, and $\mathbb{B}_\theta$ are matrices with tensor components

$$\mathbb{Q}_\theta(\mathbf{N}) \equiv \begin{bmatrix} \psi_{,\mathbf{EE}}\{\mathbf{N}\} & \psi_{,\mathbf{EK}}\{\mathbf{N}\} & \mathbf{0} \\ \psi_{,\mathbf{KE}}\{\mathbf{N}\} & \psi_{,\mathbf{KK}}\{\mathbf{N}\} & \mathbf{0} \\ -\rho_0\theta\mathbf{N}\cdot\psi_{,\theta\mathbf{E}} & -\rho_0\theta\mathbf{N}\cdot\psi_{,\theta\mathbf{E}} & \mathbf{N}\cdot\mathbf{k}(\theta) \end{bmatrix},$$

$$\mathbb{B}_\theta \equiv \begin{bmatrix} \mathbf{I} & \mathbf{0} & \mathbf{0} \\ \mathbf{0} & \gamma\mathbf{I} & \mathbf{0} \\ \mathbf{0} & \mathbf{0} & \mathbf{0} \end{bmatrix}.$$

Thus when the heat flow vector obeys Fourier's law, we reduce the problem of propagation of an acceleration wave in a thermoelastic micropolar continuum to the spectral problem (5.19). Namely, an homothermal acceleration wave exists only if (5.19) has nontrivial solutions and the eigenvalues for the problem (5.19) are real and positive.

It is easy to see that the results can be extended to the case of a nonlinear thermoconductivity law having the form $\mathbf{q} = \mathbf{q}(\theta, \mathrm{Grad}\,\theta)$. For this, it is sufficient to require that

$$\mathbf{g}\cdot\mathbf{q}_{,\mathrm{Grad}\,\theta}\cdot\mathbf{g} < 0,\quad \forall\ \mathbf{g}\neq\mathbf{0}\,.$$

In a similar manner, we can consider a more general form of the thermoconductivity law

$$\mathbf{q} = \mathbf{q}(\theta, \mathrm{Grad}\,\theta, \mathbf{E}, \mathbf{K}). \tag{5.20}$$

Here the form of corresponding matrix $\mathbf{Q}_\theta$ becomes more complex but it does not change main properties of the spectral problem.

5.4 Homocaloric (Homentropic) Acceleration Waves

By the definition of an homentropic acceleration wave, we have $[\![\eta]\!] = 0$. Let us suppose the temperature field to be continuous at $S(t)$, that is $[\![\theta]\!] = 0$. We can prove that this is valid if

$$\eta,_{\theta} \equiv -\psi,_{\theta\theta} \neq 0 \tag{5.21}$$

is true. So we suppose this to hold. A physical reason for assumption (5.21) is based on the fact that the second derivative of the free energy corresponds to the heat capacity, but a real material with zero heat capacity does not exist.

Applying Maxwell's theorem, we define a scalar thermal amplitude Θ such that

$$[\![\dot{\theta}]\!] = -V\Theta, \quad [\![\mathrm{Grad}\,\theta]\!] = \Theta\mathbf{N}. \tag{5.22}$$

Differentiating constitutive equations (5.3), we obtain

$$\left[\!\left[\dot{\mathbf{T}}_\kappa\right]\!\right] = \rho_0\psi,_{\mathbf{EE}} \cdot\cdot \left[\!\left[\dot{\mathbf{E}}\right]\!\right]^T + \rho_0\psi,_{\mathbf{EK}} \cdot\cdot \left[\!\left[\dot{\mathbf{K}}\right]\!\right]^T + \rho_0\psi,_{\mathbf{E}\theta} [\![\dot{\theta}]\!],$$

$$\left[\!\left[\dot{\mathbf{M}}_\kappa\right]\!\right] = \rho_0\psi,_{\mathbf{KE}} \cdot\cdot \left[\!\left[\dot{\mathbf{E}}\right]\!\right]^T + \rho_0\psi,_{\mathbf{KK}} \cdot\cdot \left[\!\left[\dot{\mathbf{K}}\right]\!\right]^T + \rho_0\psi,_{\mathbf{K}\theta} [\![\dot{\theta}]\!].$$

For Fourier's law (5.17), from Eq. (5.8) we get

$$\mathbf{N} \cdot \mathbf{k} \cdot [\![\mathrm{Grad}\,\theta]\!] = \mathbf{N} \cdot \mathbf{k} \cdot \mathbf{N}\Theta = 0. \tag{5.23}$$

For a more general thermoconductivity law (5.20), using $\mathbf{k} = -\mathbf{q},_{\mathrm{Grad}\theta}$, we also can get (5.23). Relation (5.23) means that for any heat conductive media with a positive definite thermoconductivity tensor, the thermal amplitude is zero. Thus, for a heat conductive media we have established that the acceleration wave should be homothermal. This result was presented by [2] for simple materials.

For heat non-conductors, $\mathbf{k} = \mathbf{0}$. We can use this assumption if we neglect the heat conductivity or consider deformation processes to be very fast. Considering a heat non-conductive media, we may obtain a homocaloric acoustic tensor that differs, in general, from $\mathbf{Q}$.

5.5 Existence of Acceleration Waves

In what follows, we will treat a heat conductive media only.

Let us consider the problem of existence of positive solutions to the spectral problem (5.19). In general, $\mathbb{Q}_\theta$ is not symmetric and $\mathbb{B}_\theta$ is not positive definite. However, it is possible to extend the *Fresnel–Hadamard–Duhem theorem* to the case of a thermoelastic micropolar continuum.

Theorem 5.2 *For any propagation directions defined by the vector* **N**, *the homothermal acoustic numbers are real.*

Proof The acoustic numbers are the squared speeds of propagation of acceleration waves. Spectral problem (5.19) splits into two problems, namely the problems (5.13) and (5.18). The components of $\mathbb{Q}$ are composed of the mixed derivatives of the free energy ψ and so $\mathbb{Q}$ is symmetric. $\mathbb{B}$ also is a symmetric matrix, moreover it is positively definite. So spectral problem (5.13) has only real-valued solutions. Let us solve problem (5.13). As $\mathbf{k}$ is a nonsingular tensor, from Eq. (5.18) we find vector **g** that is uniquely defined.

Thus the theorem is proved under the assumptions of [9], that are supplemented by the requirement that the heat conductivity tensor $\mathbf{k}$ must be positively definite. □

This theorem does not guarantee existence of an acceleration wave as problem (5.13) may have zero or negative eigenvalues. For existence of acceleration waves, all the eigenvalues of (5.13) must be positive for any **N**. Thus, we should impose some additional restrictions on the constitutive equations. We will use the strong ellipticity condition as such a constitutive restriction. For a thermoelastic micropolar media, the condition is represented by the inequality

$$\frac{d^2}{d\tau^2}\psi(\mathbf{E}+\tau\mathbf{a}'\otimes\mathbf{N},\mathbf{K}+\tau\mathbf{b}'\otimes\mathbf{N})\Big|_{\tau=0}>0,\quad \forall\mathbf{N}:|\mathbf{N}|=1,\quad \mathbf{a}'\neq\mathbf{0},\quad \mathbf{b}'\neq\mathbf{0}, \tag{5.24}$$

We establish the following theorem that is similar to the theorems in [9].

Theorem 5.3 *The condition for existence of a homothermal acceleration wave for all directions of propagations in a micropolar thermoelastic continuum is equivalent to the condition of strong ellipticity of the equilibrium equations of the continuum.*

Proof Acceleration waves exist only if all eigenvalues of spectral problem (5.19) are positive for any **N** defining the direction of wave propagation and it must hold $V^2>0$. As problem (5.19) is equivalent to two problems, (5.13) and (5.18), these properties of positiveness are valid if and only if $\mathbb{Q}$ is positively definite for any values of **N**. So by the definition of positiveness, we have

$$\boldsymbol{\xi}\cdot\mathbb{Q}(\mathbf{N})\cdot\boldsymbol{\xi}>0,\quad \forall\mathbf{N}:|\mathbf{N}|=1,\quad \boldsymbol{\xi}\neq\mathbf{0}. \tag{5.25}$$

This is an additional restriction that is imposed on the constitutive relations of the thermoelastic micropolar continuum. It is easily seen that inequality (5.25) coincides with (5.24) that completes the proof. □

Let us note that positive definiteness of $\mathbf{k}$ implies strong ellipticity of the steady-state thermoconductivity equation. Degeneration of at least one of the quantities $\mathbb{Q}$ or $\mathbf{k}$ leads to the possibility of existence of non-smooth solutions to the equilibrium equations or the steady-state thermoconductivity equation.

We wish to underline that the existence of acceleration waves in all the directions and the equivalence to the condition of strong ellipticity are local as they are defined at each point of the continuum. In case of non-homogeneous deformation this means, that the conditions can break or be valid in different parts of the medium.

5.6 Example

As an example, let us consider a quadratic form as a constitutive equation for the specific free energy. Let us assume the following relation to be valid

$$\begin{aligned}\rho_0\psi &= W_1(\mathbf{E}) + W_2(\mathbf{K}),\\ 2W_1(\mathbf{E}) &= \alpha_1 \mathrm{tr}\left(\mathbf{E}\cdot\mathbf{E}^T\right) + \alpha_2 \mathrm{tr}\,\mathbf{E}^2 + \alpha_3 \mathrm{tr}^2\mathbf{E} + \alpha_0(\theta-\theta_0)\mathrm{tr}\,\mathbf{E} + c(\theta-\theta_0)^2,\\ 2W_2(\mathbf{K}) &= \beta_1 \mathrm{tr}\left(\mathbf{K}\cdot\mathbf{K}^T\right) + \beta_2 \mathrm{tr}\,\mathbf{K}^2 + \beta_3 \mathrm{tr}^2\mathbf{K},\end{aligned} \tag{5.26}$$

where α_k, β_k $(k = 1, 2, 3)$ are elastic constants, α_0 corresponds to the thermal expansion coefficient, c is the specific heat capacity, and θ_0 is the reference temperature.

The acoustic tensor $\mathbb{Q}(\mathrm{N})$ is given by

$$\mathbb{Q}(\mathrm{N}) \equiv \begin{bmatrix} \mathbf{Q}_1(\mathrm{N}) & \mathbf{0} \\ \mathbf{0} & \mathbf{Q}_2(\mathrm{N}) \end{bmatrix}, \quad \rho_0\mathbf{Q}_1(\mathrm{N}) = W_{1,\mathbf{EE}}\{\mathrm{N}\}, \quad \rho_0\mathbf{Q}_2(\mathrm{N}) = W_{2,\mathbf{KK}}\{\mathrm{N}\}.$$

Thus, spectral problem (5.13) splits into the two problems

$$\mathbf{Q}_1(\mathrm{N})\cdot\mathbf{a}' = V^2\mathbf{a}', \qquad \mathbf{Q}_2(\mathrm{N})\cdot\mathbf{b}' = \gamma V^2\mathbf{b}'. \tag{5.27}$$

For constitutive equation (5.26), inequality (5.25) splits into two inequalities

$$\mathbf{a}\cdot\mathbf{Q}_1(\mathrm{N})\cdot\mathbf{a} > 0, \quad \mathbf{b}\cdot\mathbf{Q}_2(\mathrm{N})\cdot\mathbf{b} > 0, \quad \forall \mathrm{N} : |\mathrm{N}| = 1, \quad \mathbf{a} \neq \mathbf{0}, \quad \mathbf{b} \neq \mathbf{0}, \tag{5.28}$$

The inequalities (5.28) are equivalent to

$$\alpha_1(\mathbf{a}\cdot\mathbf{a})^2 + (\alpha_2+\alpha_3)(\mathbf{a}\cdot\mathrm{N})^2 > 0, \qquad \beta_1(\mathbf{b}\cdot\mathbf{a})^2 + (\beta_2+\beta_3)(\mathbf{b}\cdot\mathrm{N})^2 > 0,$$

that implies the following

$$\alpha_1 > 0, \quad \alpha_1 + \alpha_2 + \alpha_3 > 0, \quad \beta_1 > 0, \quad \beta_1 + \beta_2 + \beta_3 > 0. \tag{5.29}$$

If (5.29) are valid then the system of equations for a physically linear material defined by relation (5.27) are strongly elliptic for any deformations. Then the solutions of (5.27) are given by

$$V_{1,2} = \sqrt{\frac{\alpha_1}{\rho_0}}, \quad \boldsymbol{\xi}_{1,2} = (\mathbf{e}_{1,2}, \mathbf{0}), \quad V_3 = \sqrt{\frac{\alpha_1 + \alpha_2 + \alpha_3}{\rho_0}}, \quad \boldsymbol{\xi}_3 = (\mathbf{0}, \mathbf{N}), \tag{5.30}$$

$$V_{4,5} = \sqrt{\frac{\beta_1}{\gamma\rho_0}}, \quad \boldsymbol{\xi}_{4,5} = (\mathbf{e}_{4,5}, \mathbf{0}), \quad V_6 = \sqrt{\frac{\beta_1 + \beta_2 + \beta_3}{\gamma\rho_0}}, \quad \boldsymbol{\xi}_6 = (\mathbf{0}, \mathbf{N}), \tag{5.31}$$

where $\mathbf{e}_1, \mathbf{e}_2, \mathbf{e}_4, \mathbf{e}_5$ are arbitrary unit vectors in the tangential plane to $S(t)$ such that $\mathbf{e}_1 \cdot \mathbf{e}_2 = \mathbf{e}_1 \cdot \mathbf{N} = \mathbf{e}_2 \cdot \mathbf{N} = 0$, $\mathbf{e}_4 \cdot \mathbf{e}_5 = \mathbf{e}_4 \cdot \mathbf{N} = \mathbf{e}_5 \cdot \mathbf{N} = 0$.

Solutions (5.30) describe *transverse and longitudinal acceleration waves*, respectively, while (5.31) describe *transverse and longitudinal acceleration waves of microrotation*. The speeds from (5.30) coincide with the limits of the phase velocities of plane harmonic waves (acoustic waves) in linear micropolar elasticity (see, [12, 15, 17]) when the frequency of the waves tends to infinity.

Recently, engineers use widely cellular solids and metal or polymer foams [18, 19]. To describe the behavior of these complex materials, they use the model of a Cosserat continuum [20–24] and more complex models of continuum mechanics (cf., for example, [25–27]). The elastic constants of such materials show a significant temperature dependence. For example, Gibson and Ashby [19] proposed a linear approximation of Young's modulus with respect to temperature having the form $E = E_0(1 - \kappa\theta/\theta_0)$, where θ_0 is now a melting temperature for metal foams or glass temperature for polymer foams, and E_0 and κ are constants.

If one assumes more general relations

$$\alpha_i(\theta) = \alpha_i^0\left(1 - \kappa_i\frac{\theta}{\theta_0}\right), \quad \beta_i(\theta) = \beta_i^0\left(1 - \gamma_i\frac{\theta}{\theta_0}\right)$$

then inequalities (5.29) are nontrivial; they depend on temperature. In the range of temperature where condition (5.29) is not valid, one may expect singular behavior in the strain state; shear bands can arise, for example.

References

1. A.C. Eringen, E.S. Suhubi, *Elastodynamics*, vol. 1 (Academic Press, New York, 1974)
2. C. Truesdell, W. Noll, The nonlinear field theories of mechanics. in *Handbuch der Physik*, vol. III/3, ed. by S. Flügge (Springer, Berlin, 1965), pp. 1–602
3. C. Truesdell, *A First Course in Rational Continuum Mechanics* (Academic Press, New York, 1977)

4. C. Truesdell, *Rational Thermodynamics*, 2nd edn. (Springer, New York, 1984)
5. B. Straughan, *Stability and Wave Motion in Porous Media, Applied Mathematical Sciences*, vol. 165 (Springer, New York, 2008)
6. B. Straughan, *Heat Waves, Applied Mathematical Sciences*, vol. 177 (Springer, New York, 2011)
7. C.B. Kafadar, A.C. Eringen, Micropolar media—I. The classical theory. Int. J. Eng. Sci. **9**(3), 271–305 (1971)
8. G.A. Maugin, Acceleration waves in simple and linear viscoelastic micropolar materials. Int. J. Eng. Sci. **12**(2), 143–157 (1974)
9. V.A. Eremeyev, Acceleration waves in micropolar elastic media. Doklady Phys. **50**(4), 204–206 (2005)
10. V.A. Eremeyev, L.M. Zubov, On the stability of elastic bodies with couple stresses. Mech. Solids **29**(3), 172–181 (1994)
11. A. Dietsche, P. Steinmann, K. Willam, Micropolar elastoplasticity and its role in localization. Int. J. Plast. **9**(7), 813–831 (1993)
12. A.C. Eringen, *Microcontinuum Field Theory. I. Foundations and Solids* (Springer, New York, 1999)
13. V.I. Erofeev, *Wave Processes in Solids with Microstructure* (World Scientific, Singapore, 2003)
14. G.A. Maugin, *Nonlinear Waves on Elastic Crystals* (Oxford University Press, Oxford, 1999)
15. W. Nowacki, *Theory of Asymmetric Elasticity* (Pergamon-Press, Oxford, 1986)
16. A.C. Eringen, C.B. Kafadar, Polar field theories. in *Continuum Physics*, vol. IV, ed. by A.C. Eringen (Academic Press, New York, 1976), pp. 1–75
17. V.A. Pal'mov, Fundamental equations of the theory of asymmetric elasticity. J. Appl. Mech. Math. **28**(3), 496–505 (1964)
18. M.F. Ashby, A.G. Evans, N.A. Fleck, L.J. Gibson, J.W. Hutchinson, H.N.G. Wadley, *Metal Foams: A Design Guid* (Butterworth-Heinemann, Boston, 2000)
19. L.J. Gibson, M.F. Ashby, *Cellular Solids: Structure and Properties. Cambridge Solid State Science Series*, 2nd edn. (Cambridge University Press, Cambridge, 1997)
20. S. Diebels, H. Steeb, Stress and couple stress in foams. Comput. Mater. Sci. **28**(3–4), 714–722 (2003)
21. R.S. Lakes, Experimental microelasticity of two porous solids. Int. J. Solids Struct. **22**(1), 55–63 (1986)
22. R.S. Lakes, Experimental Micro Mechanics Methods for Conventional and Negative Poisson's Ratio Cellular Solids as Cosserat Continua. Trans. ASME J. Eng. Mater. Technol. **113**(1), 148–155 (1991)
23. R.S. Lakes, Experimental methods for study of Cosserat elastic solids and other generalized continua. in *Continuum Models for Materials with Micro-Structure*, ed. by H. Mühlhaus (Wiley, New York, 1995), pp. 1–22
24. H.C. Park, R.S. Lakes, Cosserat micromechanics of human bone: strain redistribution by a hydration-sensitive constituent. J. Biomech. **19**(5), 385–397 (1986)
25. T. Dillard, S. Forest, P. Ienny, Micromorphic continuum modelling of the deformation and fracture behaviour of nickel foams. Eur. J. Mech. A-Solids **25**(3), 526–549 (2006)
26. S. Forest, K.T. Duy, Generalized continua and non-homogeneous boundary conditions in homogenisation methods. ZAMM **91**(2), 90–109 (2011)
27. P. Neff, S. Forest, A geometrically exact micromorphic model for elastic metallic foams accounting for affine microstructure. Modelling, existence of minimizers, identification of moduli and computational results. J. Elast. **87**(2–3), 239–276 (2007)

Appendix A
Elements of Tensor Analysis

In this appendix we present the formulae of tensor calculus that are in the book. We present the theory in space $\mathbb{R}^3$. For more details, one can recommend for example [1].

A.1 Vectors

The representation of a vector relates with a coordinate system and corresponding basis in $\mathbb{R}^3$. Let $\mathbf{e}_1, \mathbf{e}_2$, and $\mathbf{e}_3$ be a basis in $\mathbb{R}^3$. Then $\mathbf{a}$ can be spanned in the form

$$\mathbf{a} = \sum_{i=1}^{3} a^i \mathbf{e}_i, \qquad \text{(A.1)}$$

where a^i are the components of $\mathbf{a}$ in basis $\mathbf{e}_i$. In what follows we will use the *Einstein rule of summation* over repeated indices so the last formula becomes $\mathbf{a} = a^i \mathbf{e}_i$. We will denote $\mathbf{i}_k$, $k = 1, 2, 3$ a canonic orthonormal basis in $\mathbb{R}^3$ that relates with the Cartesian coordinate system.

In space $\mathbb{R}^3$ we introduce the *dot product*. For $\mathbf{a}, \mathbf{b} \in \mathbb{R}^3$ it is defined by

$$\mathbf{a}\cdot\mathbf{b} = |\mathbf{a}||\mathbf{b}| \cos\theta,$$

where $|\mathbf{a}|$ and $|\mathbf{b}|$ are absolute values of $\mathbf{a}$ and $\mathbf{b}$ respectively and θ is the least angle between $\mathbf{a}$ and $\mathbf{b}$. In canonic basis $\mathbf{i}_1, \mathbf{i}_2, \mathbf{i}_3$,

$$\mathbf{a} = a_1\mathbf{i}_1 + a_2\mathbf{i}_2 + a_3\mathbf{i}_3, \qquad \mathbf{b} = b_1\mathbf{i}_1 + b_2\mathbf{i}_2 + b_3\mathbf{i}_3,$$

and for the dot product we get

$$\mathbf{a}\cdot\mathbf{b} = a_1b_1 + a_2b_2 + a_3b_3.$$

V. A. Eremeyev et al., *Foundations of Micropolar Mechanics*,
SpringerBriefs in Continuum Mechanics,
DOI: 10.1007/978-3-642-28353-6, © The Author(s) 2013

For $\mathbf{a}, \mathbf{b} \in \mathbb{R}^3$ we also introduce the *cross product* as follows

$$\mathbf{a} \times \mathbf{b} = \begin{vmatrix} \mathbf{i}_1 & \mathbf{i}_2 & \mathbf{i}_3 \\ a_1 & a_2 & a_3 \\ b_1 & b_2 & b_3 \end{vmatrix} \equiv (a_2b_3 - a_3b_2)\mathbf{i}_1 - (a_1b_3 - a_3b_1)\mathbf{i}_2 + (a_1b_2 - a_2b_1)\mathbf{i}_3.$$

The result of the cross product of **a** by **b** is a vector that is orthogonal to **a** and **b**, its absolute value is

$$|\mathbf{a} \times \mathbf{b}| = |\mathbf{a}||\mathbf{b}| \sin\theta.$$

For the cross product we have $\mathbf{a} \times \mathbf{b} = -\mathbf{b} \times \mathbf{a}$.

Calculating the cross product in a canonic basis, one can use the following formulae

$$\mathbf{i}_1 \times \mathbf{i}_2 = \mathbf{i}_3, \quad \mathbf{i}_2 \times \mathbf{i}_3 = \mathbf{i}_1, \quad \mathbf{i}_3 \times \mathbf{i}_1 = \mathbf{i}_2.$$

Let $\mathbf{e}_i$ be an arbitrary basis in $\mathbb{R}^3$. We denote by $\mathbf{e}^j$ the dual basis to $\mathbf{e}_i$ that is given by the formulae

$$\mathbf{e}_j \cdot \mathbf{e}^i = \delta^i_j \qquad (i,j = 1,2,3), \tag{A.2}$$

where

$$\delta^i_j = \begin{cases} 1, & j = i, \\ 0, & j \neq i \end{cases}$$

is *Kronecker's symbol*. It is seen that the dual to a canonic basis coincides with the principle basis, that is $\mathbf{i}^k = \mathbf{i}_k, k = 1,2,3$.

Any vector **a** can be presented with use of the principle basis or its dual basis:

$$\mathbf{a} = a^i\mathbf{e}_i = a_i\mathbf{e}^i, \quad a_i = \mathbf{a} \cdot \mathbf{e}_i, \quad a^i = \mathbf{a} \cdot \mathbf{e}^i.$$

a_i are called *covariant components* of **a** whereas a^i are its *contravariant components*.

The quantities $g_{ij} = \mathbf{e}_i \cdot \mathbf{e}_j$ and $g^{ij} = \mathbf{e}^i \cdot \mathbf{e}^j$ are called *metric coefficients*. The sets g_{ij} and g^{ij} constitute 3×3 matrices. It can be shown that (g^{ij}) is the inverse matrix to (g_{ij}) as

$$g_{ij}g^{jk} = \delta^k_i.$$

Using the metric coefficients we can represent the dot product of **a** by **b** by the formulae

$$\mathbf{a} \cdot \mathbf{b} = g_{ij}a^ib^j = g^{ij}a_ib_j = a_ib^i = a^ib_j.$$

A.2 Tensors

In what follows $\mathbb{V}$ with possible indices will denote $\mathbb{R}^3$ or its subspaces. The principle notion of tensor calculus is the *tensor product*. Let vectors $\mathbf{a} \in \mathbb{V}_1$, $\mathbf{b} \in \mathbb{V}_2$. The tensor product of $\mathbf{a}$ by $\mathbf{b}$ is the ordered pair $\mathbf{a} \otimes \mathbf{b}$ called the *dyad*. The dyads have the following properties:

$$\begin{aligned}
(\lambda \mathbf{a}) \otimes \mathbf{b} &= \mathbf{a} \otimes (\lambda \mathbf{b}) = \lambda(\mathbf{a} \otimes \mathbf{b}),\\
(\mathbf{a}+\mathbf{b}) \otimes \mathbf{c} &= \mathbf{a} \otimes \mathbf{c} + \mathbf{b} \otimes \mathbf{c},\\
\mathbf{a} \otimes (\mathbf{b}+\mathbf{c}) &= \mathbf{a} \otimes \mathbf{b} + \mathbf{a} \otimes \mathbf{c},
\end{aligned} \tag{A.3}$$

where $\lambda \in \mathbb{R}$. We underline that the dyad is an ordered pair that means that for $\mathbf{a}$ that is not proportional to $\mathbf{b}$ it follows $\mathbf{a} \otimes \mathbf{b} \neq \mathbf{b} \otimes \mathbf{a}$.

The tensor product operation applied to all possible vectors of $\mathbb{V}_1$ and $\mathbb{V}_2$ is called the tensor product of $\mathbb{V}_1$ by $\mathbb{V}_2$, it is denoted by $\mathbb{T}_2 = \mathbb{V}_1 \otimes \mathbb{V}_2$. Subscript 2 in $\mathbb{T}_2$ evidences that in the product there participate two vectorial spaces. In a similar manner we can introduce the tensor product of n vectorial spaces

$$\mathbb{T}_n = \mathbb{V}_1 \otimes \mathbb{V}_2 \otimes \cdots \otimes \mathbb{V}_n.$$

In applications, they normally use the tensor products of two, three and four vectorial spaces, that is $\mathbb{T}_2$, $\mathbb{T}_3$, and $\mathbb{T}_4$. The elements of $\mathbb{T}_n$ are called the tensors of nth order. The vectors of $\mathbb{V}$ can be formally considered as the first-order tensors, that is $\mathbb{V} \equiv \mathbb{T}_1$, whereas the scalars are the zero-order tensors.

In general, vector spaces that are used to generate the tensor products can have different dimensions. For example, in the shell theory they use the tensor product of a two dimensional space by a three dimensional. Here we will restrict ourselves to the case when $\mathbb{V}_1 = \mathbb{V}_2 = \mathbb{V}$.

Let $\mathbf{e}_i$ be a basis of $\mathbb{V}$. Expanding the vectors over the basis and using properties (A.3), we represent $\mathbf{a} \otimes \mathbf{b}$ in the form

$$\mathbf{a} \otimes \mathbf{b} = a^i \mathbf{e}_i \otimes b^j \mathbf{e}_j = a^i b^j \mathbf{e}_i \otimes \mathbf{e}_j.$$

In $\mathbb{R}^3 \otimes \mathbb{R}^3$ there are nine dyads $\mathbf{e}_i \otimes \mathbf{e}_j$ that constitute a basis of $\mathbb{R}^3 \otimes \mathbb{R}^3$. Again, the elements of $\mathbb{R}^3 \otimes \mathbb{R}^3$ are called second-order tensors. Each second-order tensor A is uniquely expanded as a linear combination of the basis dyads

$$\mathsf{A} = a^{ij} \mathbf{e}_i \otimes \mathbf{e}_j.$$

Instead of $\mathbf{e}_i \otimes \mathbf{e}_j$ we can use dyads $\mathbf{e}^i \otimes \mathbf{e}^j$, $\mathbf{e}_i \otimes \mathbf{e}^j$ or $\mathbf{e}^i \otimes \mathbf{e}_j$. So the tensor A has other representations

$$\mathsf{A} = a_{ij} \mathbf{e}^i \otimes \mathbf{e}^j = a_i^{\cdot j} \mathbf{e}^i \otimes \mathbf{e}_j = a^i_{\cdot j} \mathbf{e}_i \otimes \mathbf{e}^j$$

as well. Quantities a_{ij} are called the covariant components of A, a^{ij} are its contravariant components and $a^i_{\cdot j}$ and $a_i^{\cdot j}$ are the *mixed components* of A. If we change the basis of $\mathbb{V}$ then the components of the tensor change by certain rules. So we can the tensor as the matrix (matrices) of its components supplied by the rules of the component transformations under the change of the basis.

Let us consider principle operation with second-order tensors. As elements of a vector space, we can add and subtract tensors as well multiply them by scalars. Besides we introduce the following operations that mimic the operations known for square matrices.

1. *Dot product.* The dot product of dyad $\mathbf{a} \otimes \mathbf{b}$ and $\mathbf{c} \otimes \mathbf{d}$ is

$$(\mathbf{a} \otimes \mathbf{b}) \cdot (\mathbf{c} \otimes \mathbf{d}) = (\mathbf{b} \cdot \mathbf{c}) \mathbf{a} \otimes \mathbf{d}.$$

The result is a dyad (a second-order tensor if we dot multiply two second-order tensors).

2. *Double dot product.*

$$(\mathbf{a} \otimes \mathbf{b}) \cdot\cdot (\mathbf{c} \otimes \mathbf{d}) = (\mathbf{b} \cdot \mathbf{c})(\mathbf{a} \cdot \mathbf{d}).$$

The result is a scalar quantity.

3. *Inner product.*

$$(\mathbf{a} \otimes \mathbf{b}) \bullet (\mathbf{c} \otimes \mathbf{d}) = (\mathbf{a} \cdot \mathbf{c})(\mathbf{b} \cdot \mathbf{d}).$$

The result is a scalar quantity again. Unlike the double dot product, the inner product has all the properties of an inner product in a vector space.

4. *Cross product.*

$$(\mathbf{a} \otimes \mathbf{b}) \times (\mathbf{c} \otimes \mathbf{d}) = \mathbf{a} \otimes (\mathbf{b} \times \mathbf{c}) \otimes \mathbf{d}).$$

The result is a triad, a third-order tensor.

By linearity, these operation can be extended to tensors of any order. For example, the dot product of $\mathbf{A}$ by $\mathbf{B}$ is

$$\mathbf{A} \cdot \mathbf{B} = \left(a^{ij} \mathbf{e}_i \otimes \mathbf{e}_j\right) \cdot \left(b^{mn} \mathbf{e}_m \otimes \mathbf{e}_n\right) = a^{ij} b^{mn} g_{jm} \mathbf{e}_i \otimes \mathbf{e}_n.$$

The inner product of second-order tensors induces the Euclidean norm in the space of second-order tensors $\|\mathbf{A}\| = \sqrt{\mathbf{A} \bullet \mathbf{A}}$.

A.3 Second-Order Tensors

Let us consider the second-order tensors that are most frequently used in applications. A second-order tensor $\mathbf{A}$ can be considered as a linear operator acting in vector space $\mathbb{V}$ by the rule

$$\mathbf{y} = \mathbf{A} \cdot \mathbf{y}, \quad \mathbf{x}, \mathbf{y} \in \mathbb{V}.$$

Examples of second-order tensors are the unit tensor $\mathbf{I}$ and the zero tensor $\mathbf{0}$, for which

$$\mathbf{I}\cdot\mathbf{x} = \mathbf{x}, \quad \mathbf{0}\cdot\mathbf{x} = \mathbf{0}, \quad \forall\mathbf{x} \in \mathbb{V}$$

The *inverse* to $\mathbf{A}$ is tensor $\mathbf{A}^{-1}$ such that

$$\mathbf{A}\cdot\mathbf{A}^{-1} = \mathbf{I}.$$

A tensor for which there exists the inverse tensor is called non-degenerated or non-singular.

Transposition of $\mathbf{A}$ denoted by $\mathbf{A}^T$, is the tensor in which the basis vector in the dyads are transposed, that is

$$\mathbf{A}^T \equiv \left(a^{ij}\mathbf{e}_i \otimes \mathbf{e}_j\right)^T = a^{ij}\mathbf{e}_j \otimes \mathbf{e}_i.$$

The transposition of a dyad is $(\mathbf{a} \otimes \mathbf{b})^T = \mathbf{b} \otimes \mathbf{a}$.

Important characteristics of a tensor are its *trace* and its *determinant* denoted by $\operatorname{tr}\mathbf{A}$ and $\det\mathbf{A}$, respectively. The trace is the sum of the mixed diagonal components of the tensor

$$\operatorname{tr}\mathbf{A} = a_i^{\cdot i} = a^i_{\cdot i}$$

whereas $\det\mathbf{A}$ is the determinant of the matrix of the mixed tensorial components

$$\det\mathbf{A} = |a_i^{\cdot j}| = |a^i_{\cdot j}|.$$

In particular for a dyad, $\operatorname{tr}(\mathbf{a} \otimes \mathbf{b}) = \mathbf{a}\cdot\mathbf{b}$ and the determinant of any dyad is zero. The following properties are valid:

$$\operatorname{tr}(\mathbf{A}) = \operatorname{tr}(\mathbf{A}^T), \quad \operatorname{tr}(\mathbf{A}\cdot\mathbf{B}) = \operatorname{tr}(\mathbf{B}\cdot\mathbf{A}),$$

$$\det(\mathbf{A}) = \det(\mathbf{A}^T), \quad \det(\mathbf{A}\cdot\mathbf{B}) = \det(\mathbf{A})\det(\mathbf{B}).$$

There hold the formulae

$$\mathbf{A}\cdot\cdot\mathbf{B} = \operatorname{tr}(\mathbf{A}\cdot\mathbf{B}), \quad \mathbf{A} \bullet \mathbf{B} = \operatorname{tr}(\mathbf{A}\cdot\mathbf{B}^T).$$

For the second-order tensor $\mathbf{X}$ one can introduce the vector invariant $\mathbf{X}_\times$ by the formula

$$\mathbf{X}_\times = \left(X_{mn}\mathbf{e}^m \otimes \mathbf{e}^n\right)_\times \triangleq X_{mn}\mathbf{e}^m \times \mathbf{e}^n \tag{A.4}$$

for any base vectors $\mathbf{e}^k$. In particular, for a dyad $\mathbf{a} \otimes \mathbf{b}$ we have $(\mathbf{a} \otimes \mathbf{b})_\times = \mathbf{a} \times \mathbf{b}$.

A second-order tensor is called *symmetric* if $\mathbf{A} = \mathbf{A}^T$. A second-order tensor is called *skew-symmetric* if $\mathbf{A} = -\mathbf{A}^T$. Any skew-symmetric tensor $\mathbf{A}$ can be represented by the vector $\mathbf{a}$ as follows

$$\mathbf{A} = \mathbf{a} \times \mathbf{I}, \tag{A.5}$$

while the corresponding $\mathbf{A}$ axial vector $\mathbf{a}$ is given by

$$\mathbf{a} = -\frac{1}{2}\mathbf{A}_{\times}. \tag{A.6}$$

An arbitrary second-order tensor can be uniquely presented as the sum of a symmetric and skew-symmetric tensors

$$\mathbf{A} = \mathbf{S} + \boldsymbol{\Omega}, \quad \mathbf{S} = \mathbf{S}^T, \quad \boldsymbol{\Omega} = -\boldsymbol{\Omega}^T,$$

that are

$$\mathbf{S} = \frac{1}{2}(\mathbf{A} + \mathbf{A}^T), \quad \boldsymbol{\Omega} = \frac{1}{2}(\mathbf{A} - \mathbf{A}^T).$$

The second-order tensor $\mathbf{Q}$ is called *orthogonal* if it satisfies the relation

$$\mathbf{Q}\cdot\mathbf{Q}^T = \mathbf{Q}^T\cdot\mathbf{Q} = \mathbf{I}.$$

In other words, the inverse tensor $\mathbf{Q}^{-1}$ of $\mathbf{Q}$ coincide with its transposed, $\mathbf{Q}^{-1} = \mathbf{Q}^T$.

If $\det\mathbf{Q} = 1$ the orthogonal tensor $\mathbf{Q}$ is called *proper orthogonal* or *rotation tensor*. The proper orthogonal tensor $\mathbf{Q}$ describing the rotation about axis $\mathbf{e}$ for angle φ can be represented by *Gibbs' formula*

$$\mathbf{Q} = (\mathbf{I} - \mathbf{e}\otimes\mathbf{e})\cos\varphi + \mathbf{e}\otimes\mathbf{e} - \mathbf{e}\times\mathbf{I}\sin\varphi, \tag{A.7}$$

where φ is the rotation angle about the axis with the unit vector $\mathbf{e}$. Introducing the *finite rotation vector* $\boldsymbol{\theta} = 2\mathbf{e}\tan\varphi/2$ and using the formulae

$$\cos\varphi = \frac{1-\tan^2\varphi/2}{1+\tan^2\varphi/2}, \quad \sin\varphi = \frac{2\tan\varphi/2}{1+\tan^2\varphi/2},$$

we obtain the representation of $\mathbf{Q}$ in the form that does not contain trigonometric functions

$$\mathbf{Q} = \frac{1}{(4+\theta^2)}\left[(4-\theta^2)\mathbf{I} + 2\boldsymbol{\theta}\otimes\boldsymbol{\theta} - 4\mathbf{I}\times\boldsymbol{\theta}\right], \quad \theta^2 = \boldsymbol{\theta}\cdot\boldsymbol{\theta}. \tag{A.8}$$

The other known vectorial parameterizations of an orthogonal tensor are presented in [2]. By Eq. (A.8), a proper orthogonal tensor $\mathbf{Q}$ defines uniquely the vector $\boldsymbol{\theta}$

$$\boldsymbol{\theta} = 2(1+\operatorname{tr}\mathbf{Q})^{-1}\mathbf{Q}_{\times}. \tag{A.9}$$

The second-order tensor $\mathbf{U}$ is called *unimodular* if it satisfies the relation

$$\det\mathbf{U} = \pm 1.$$

The set of all unimodular tensors constitutes the group with regard to multiplication which we denote *Unim*. The set of all orthogonal tensors and the set of all rotations tensors constitute the groups *Orth* and *Orth*$^+$, respectively. Second-order tensors can be also considered as elements of the linear group *Lin* that is the group with regard to addition.

A nontrivial solution $\mathbf{x}$ to the equation

$$\mathbf{A}\cdot\mathbf{x} = \lambda\mathbf{x} \tag{A.10}$$

at some value λ is called the *eigenvector* whereas λ is the *eigenvalue* of tensor $\mathbf{A}$; $(\lambda, \mathbf{x})$, $\mathbf{x} \neq \mathbf{0}$ constitute the eigenvalue. From (A.10) it is seen that $a\mathbf{x}, a \in \mathbb{R}$ is also an eigenvector so we can take a such that $\|\mathbf{x}\| = 1$. We rewrite (A.10) in the form

$$(\mathbf{A} - \lambda\mathbf{I})\cdot\mathbf{x} = \mathbf{0}.$$

It follows the eigenvalues of $\mathbf{A}$ are solutions of the characteristic equation

$$\det(\mathbf{A} - \lambda\mathbf{I}) = 0$$

that can be written as follows

$$-\lambda^3 + I_1(\mathbf{A})\lambda^2 - I_2(\mathbf{A})\lambda + I_3(\mathbf{A}) = 0. \tag{A.11}$$

Equation (A.11) has no more than three different roots λ_1, λ_2, λ_3. The coefficients $I_1(\mathbf{A})$, $I_2(\mathbf{A})$, and $I_3(\mathbf{A})$ are called *first, second, and third principal invariants* of $\mathbf{A}$, respectively. Through the components of $\mathbf{A}$, they have the form

$$I_1(\mathbf{A}) = \operatorname{tr}\mathbf{A}, \quad I_2(\mathbf{A}) = \frac{1}{2}\left(\operatorname{tr}^2\mathbf{A} - \operatorname{tr}\mathbf{A}^2\right), \quad I_3(\mathbf{A}) = \det\mathbf{A}.$$

By Vieta's formulae, the invariants can be represented through the eigenvalues λ_1, λ_2, λ_3:

$$I_1(\mathbf{A}) = \lambda_1 + \lambda_2 + \lambda_3, \quad I_2(\mathbf{A}) = \lambda_1\lambda_2 + \lambda_1\lambda_3 + \lambda_2\lambda_3, \quad I_3(\mathbf{A}) = \lambda_1\lambda_2\lambda_3.$$

As the tensors can be treated as linear operators in a finite dimensional space, they have all the properties of linear operators and their matrix representations. In particular, a symmetric tensor $\mathbf{A}$ has three real eigenvalues that can be repeated, besides we always can choose three orthonormal eigenvectors.

For symmetric tensor $\mathbf{A}$, there is the spectral decomposition

$$\mathbf{A} = \lambda_1\mathbf{e}_1 \otimes \mathbf{e}_2 + \lambda_2\mathbf{e}_2 \otimes \mathbf{e}_2 + \lambda_3\mathbf{e}_3 \otimes \mathbf{e}_3,$$

where λ_k and $\mathbf{e}_k$, $k = 1, 2, 3$ are the eigenvalues and eigenvectors of $\mathbf{A}$, respectively.

For a second-order tensor, Cayley–Hamilton theorem holds

Theorem A.1 (Cayley–Hamilton) *Tensor* $\mathbf{A}$ *satisfies its characteristic equation*:

$$-\mathbf{A}^3 + I_1(\mathbf{A})\mathbf{A}^2 - I_2(\mathbf{A})\mathbf{A} + I_3(\mathbf{A})\mathbf{I} = \mathbf{0}. \tag{A.12}$$

By Cayley–Hamilton theorem, we can express any integer power of $\mathbf{A}$ through its lower powers and the invariants. If $\mathbf{A}$ is non-degenerated than

$$\mathbf{A}^{-1} = \frac{1}{I_3(\mathbf{A})}\left[\mathbf{A}^2 - I_1(\mathbf{A})\mathbf{A} + I_2(\mathbf{A})\mathbf{I}\right].$$

A.4 Higher Order Tensors

Examples of third- and forth-order tensors are the tensor products of three and four vectors:

$$\mathbf{a} \otimes \mathbf{b} \otimes \mathbf{c}, \quad \mathbf{a} \otimes \mathbf{b} \otimes \mathbf{c} \otimes \mathbf{d}.$$

A third-order tensor $\mathbb{A}$ can be treated as a linear operator acting from $\mathbb{V}$ to $\mathbb{T}_2$ by the rule

$$\mathbf{Y} = \mathbb{A}\cdot\mathbf{x}.$$

It also can be treated as a linear operator acting from $\mathbb{T}_2$ to $\mathbb{V}$: $\mathbf{y} = \mathbb{A} \bullet \mathbf{X}$. Similarly, a forth order tensor $\mathbb{C}$ can be treated as a linear operator from $\mathbb{T}_2$ to $\mathbb{T}_2$:

$$\mathbf{Y} = \mathbb{C} \bullet \mathbf{X}.$$

An important example of a third-order tensor is the *permutation tensor* that is also called *Levi–Civita's tensor*

$$\mathcal{E} = -\mathbf{I} \times \mathbf{I}$$

It should be noted that $\mathcal{E}$ is a pseudo-tensor. The difference between tensors and pseudo-tensors will be considered in Sect. A.6. In the component representation, $\mathcal{E}$ is

$$\mathcal{E} = \varepsilon_{ijk}\mathbf{e}^i \otimes \mathbf{e}^j \otimes \mathbf{e}^k.$$

Here the "permutation symbol" ε_{ijk} is given by $\varepsilon_{ijk} = (\mathbf{e}_j \times \mathbf{e}_k)\cdot\mathbf{e}_i$.

As a non-trivial example, we consider the "unit" forth order tensor $\mathbb{I}$, that is

$$\mathbb{I} \bullet \mathbf{X} = \mathbf{X}, \quad \forall\mathbf{X} \in \mathbb{T}_2.$$

Its component representation is

$$\mathbb{I} = \mathbf{e}_k \otimes \mathbf{e}_m \otimes \mathbf{e}^k \otimes \mathbf{e}^m.$$

Examples of forth order tensors are the tensor of elastic moduli, Riemann curvature tensor, etc.

A.5 Basis Transformation

Coordinate-free representation of tensors allow us to consider various bases in space. Let us consider the rules for the component change when we change the space bases. Let $\mathbf{e}_i$ and $\tilde{\mathbf{e}}_j$, $i,j = 1,2,3$, be two bases in $\mathbb{R}^3$ that are related by the formulae

$$\mathbf{e}_i = \sum_{j=1}^{3} A_i^j \tilde{\mathbf{e}}_j, \tag{A.13}$$

where A_i^j is a non-degenerated matrix. Then

$$\mathbf{a} = \sum_{i=1}^{3} a^i \sum_{j=1}^{3} A_i^j \tilde{\mathbf{e}}_j = \sum_{j=1}^{3} \tilde{\mathbf{e}}_j \sum_{i=1}^{3} A_i^j a^i.$$

So in the new basis $\tilde{\mathbf{e}}_j$ we have

$$\mathbf{a} = \sum_{j=1}^{3} \tilde{a}^j \tilde{\mathbf{e}}_j \equiv \tilde{a}^j \tilde{\mathbf{e}}_j, \quad \tilde{a}^j = \sum_{i=1}^{3} A_i^j x^i \equiv A_i^j a^i.$$

The dual bases $\mathbf{e}^i$ and $\tilde{\mathbf{e}}^j$ are also related by a formula analogous to (A.13) with matrix B_j^i

$$\mathbf{e}^i = \sum_{j=1}^{3} B_j^i \tilde{\mathbf{e}}^j. \tag{A.14}$$

Using identity $\tilde{\mathbf{e}}_m \cdot \tilde{\mathbf{e}}^n = \delta_m^n$, we can show that B_j^i is the inverse to A_i^j, that is

$$B_k^j A_i^k = \delta_i^j.$$

We also can show that vector $\mathbf{a}$ in basis $\tilde{\mathbf{e}}^j$ is determined by the formulae

$$\mathbf{a} = \tilde{a}_j \tilde{\mathbf{e}}^j, \quad \tilde{a}_j = B_j^i a_i.$$

So we have the formulae relating "old" and "new" components of $\mathbf{a}$:

$$a_i = A_i^j \tilde{a}_j, \quad a^i = B_j^i \tilde{a}^j. \tag{A.15}$$

These formulae demonstrate that the covariant components obey the transformation rules for the main basis (A.13) whereas the contravariant components obey the transformation rules for the dual basis (A.14).

We derive similar transformation formulae for tensors. For example, for a second-order tensor $\mathbf{A}$ we have

$$\mathbf{A} = a^{ij} \mathbf{e}_i \otimes \mathbf{e}_j = a_{kl} \mathbf{e}^k \otimes \mathbf{e}^l = a_i^{\cdot j} \mathbf{e}^i \otimes \mathbf{e}_j = a_{\cdot l}^{k} \mathbf{e}_k \otimes \mathbf{e}^l$$
$$= \tilde{a}^{ij} \tilde{\mathbf{e}}_i \otimes \tilde{\mathbf{e}}_j = \tilde{a}_{kl} \tilde{\mathbf{e}}^k \otimes \tilde{\mathbf{e}}^l = \tilde{a}_i^{\cdot j} \tilde{\mathbf{e}}^i \otimes \tilde{\mathbf{e}}_j = \tilde{a}_{\cdot l}^{k} \tilde{\mathbf{e}}_k \otimes \tilde{\mathbf{e}}^l,$$

where

$$\tilde{a}^{ij} = A^i_k A^j_l a^{kl}, \qquad a^{ij} = B^i_k B^j_l \tilde{a}^{kl},$$
$$\tilde{a}_{ij} = B^k_i B^l_j a_{kl}, \qquad a_{ij} = A^k_i A^l_j \tilde{a}_{kl},$$
$$\tilde{a}^i_{\cdot j} = A^i_k B^l_j a^k_{\cdot l}, \qquad a^i_{\cdot j} = B^i_k A^l_j \tilde{a}^k_{\cdot l},$$
$$\tilde{a}^{\cdot j}_i = B^k_i A^j_l a^{\cdot l}_k, \qquad a^{\cdot j}_i = A^k_i B^j_l \tilde{a}^{\cdot l}_k.$$

A.6 Polar and Axial Vectors and Tensors

The transformation rules for the vector components (A.15) and tensor components (A.16) determine the vectors and tensors as the quantities that are independent of the choice of the basis and the coordinate system. However, there are quantities for which the component transformation formulae differ from (A.15) or (A.16) only in the sign that depends on the coordinate system orientation, see [3, 4, 5]. For example instead of (A.15), for the component transformation we can find the relation

$$a_i = \pm A^j_i \tilde{a}_j, \quad a^i = \pm B^i_j \tilde{a}^j, \tag{A.17}$$

where sign "+" is taken when the determinant of matrix A^j_i is positive and "−" is for the negative determinant. The determinant of A^j_i is negative when the transformation is from a coordinate system with some orientation to a system with the opposite orientation. In (A.17) we choose "+" when the transformation is between the systems of equal orientation and "−" in case of opposite orientations of the systems.

Definition A.1 The vectors whose components satisfy (A.17) are called axial where as the vectors that are transformed by the rules (A.15) are called polar.

A polar vector is usually represented by a directed segment (an arrow) whereas an axial vector is represented by a line segment with the direction is shown as the rotation, see Fig. A.1. The difference in the types of the vectors on Fig. A.1 reveals when we apply the mirror reflection to the vectors with respect to plane that is perpendicular to their segment representation. After the reflection a polar vector changes its direction to the opposite one whereas the graphic representation of an axial vector does not change. Thus transformation of the mirror reflection changes the sign of a polar vector and does not change an axial vector. Examples of polar vectors are vectors of linear velocity, force, momentum, and electric field. Such vectors as angular velocity, couple, moment of momentum, magnetic field are axial. It is important to note that the cross product of two polar vectors is an axial vector.

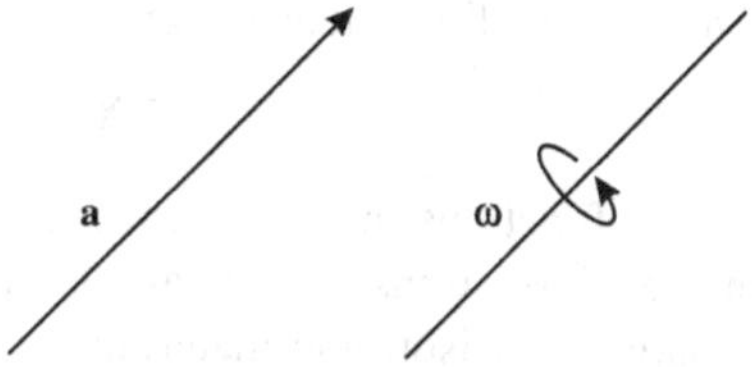

Fig. A.1 Polar vector **a** and axial vector ω vector

As well as axial and polar vectors, there exist polar and axial tensors. For the components of an axial second-order tensor, the transformation formulae are

$$\tilde{a}^{ij} = \pm A^i_k A^j_l a^{kl}, \tag{A.18}$$

where the choice in "±" is determined by the sign of the determinant of A^j_i. An example of an axial second-order tensor is the skew tensor $\mathbf{\Omega} = \mathbf{a} \times \mathbf{I}$ whose polar vector is $\mathbf{a}$. An example of the axial third-order tensor is the permutation tensor $\mathcal{E}$. The stress tensor and stretch tensor are the polar tensors, and the couple stress tensor and wryness tensors are axial. In the literature, axial vectors and tensors are also called *pseudovectors* and *pseudotensors*.

Sometimes the difference between axial and polar quantities does not affect mechanical or physical theories, so they do not distinguish them. However sometimes the nature of axial vectors and tensors becomes essential, see [3, 5]. In micropolar mechanics we should distinguish between axial and polar vectors and tensors.

A.7 Tensor Functions

In applications we meet tensor-valued functions that depend on tensors between tensorial quantities that we will call tensor functions. For tensor functions, we know some representation theorems. In particular, there are the representation theorems for the linear function.

Definition A.2 $\mathbb{F}(\mathbb{X})$ of argument $\mathbb{T}_p$ with values in $\mathbb{T}_q$ is a linear function if for all $\mathbb{X}, \mathbb{Y} \in \mathbb{T}_p$ and $\alpha, \beta \in \mathbb{R}$ there holds

$$\mathbb{F}(\alpha\mathbb{X} + \beta\mathbb{Y}) = \alpha\mathbb{F}(\mathbb{X}) + \beta\mathbb{F}(\mathbb{Y}).$$

Theorem A.2 *Let $\mathbb{F}$ be a linear function from $\mathbb{T}_p$ to $\mathbb{T}_q$. Then $\mathbb{F}(\mathbb{X}) = \mathbb{C} \bullet \mathbb{X}$, where tensor $\mathbb{C} \in \mathbb{T}_{p+q}$ is uniquely determined by $\mathbb{F}$.*

In particular, a linear vector function of vector $\mathbf{x}$ takes the form of the dot product of tensor $\mathbf{A}$ by $\mathbf{x}$:

$$\mathbf{f}(\mathbf{x}) = \mathbf{A}\cdot\mathbf{x}, \quad \mathbf{A} \in \mathbb{T}_2.$$

An arbitrary linear tensor function of a second-order tensor $\mathbf{X}$ takes the form

$$\mathbf{F}(\mathbf{X}) = \mathbb{C} \bullet \mathbf{X}, \quad \mathbb{C} \in \mathbb{T}_4.$$

The functions in nonlinear continuum mechanics are invariant under some sets of transformations. For example, isotropic functions are used for constitutive equations for isotropic materials.

Definition A.3 Scalar function f, vectorial function $\mathbf{f}$, and tensorial function $\mathbf{F}$ depending on n vectors $\mathbf{x}_i$ and m second-order tenors $\mathbf{X}_j$ are isotropic if they satisfy the following relations

$$\begin{aligned}
&f(\mathbf{Q}^T\cdot\mathbf{x}_1,\ldots,\mathbf{Q}^T\cdot\mathbf{x}_n,\mathbf{Q}^T\cdot\mathbf{X}_1\cdot\mathbf{Q},\ldots,\mathbf{Q}^T\cdot\mathbf{X}_m\cdot\mathbf{Q}) = f(\mathbf{x}_1,\ldots,\mathbf{x}_n,\mathbf{X}_1,\ldots,\mathbf{X}_m),\\
&\mathbf{f}(\mathbf{Q}^T\cdot\mathbf{x}_1,\ldots,\mathbf{Q}^T\cdot\mathbf{x}_n,\mathbf{Q}^T\cdot\mathbf{X}_1\cdot\mathbf{Q},\ldots,\mathbf{Q}^T\cdot\mathbf{X}_m\cdot\mathbf{Q}) = \mathbf{Q}^T\cdot\mathbf{f}(\mathbf{x}_1,\ldots,\mathbf{x}_n,\mathbf{X}_1,\ldots,\mathbf{X}_m),\\
&\mathbf{F}(\mathbf{Q}^T\cdot\mathbf{x}_1,\ldots,\mathbf{Q}^T\cdot\mathbf{x}_n,\mathbf{Q}^T\cdot\mathbf{X}_1\cdot\mathbf{Q},\ldots,\mathbf{Q}^T\cdot\mathbf{X}_m\cdot\mathbf{Q})\\
&= \mathbf{Q}^T\cdot\mathbf{F}(\mathbf{Q}^T\cdot\mathbf{x}_1,\ldots,\mathbf{Q}^T\cdot\mathbf{x}_n,\mathbf{Q}^T\cdot\mathbf{X}_1\cdot\mathbf{Q},\ldots,\mathbf{Q}^T\cdot\mathbf{X}_m\cdot\mathbf{Q})\cdot\mathbf{Q}
\end{aligned}$$

respectively for any orthogonal tensor $\mathbf{Q}$.

If Definition A.3 is valid for proper orthogonal tensors $\mathbf{Q}$ only, the functions are called *hemitropic*.

A linear isotropic tensor function $\mathbf{F}(\mathbf{X})$ from $\mathbb{T}_2$ to $\mathbb{T}_2$ takes the form

$$\mathbf{F}(\mathbf{X}) = \alpha\mathbf{I}\,\mathrm{tr}\,\mathbf{X} + \beta\mathbf{X} + \gamma\mathbf{X}^T, \quad \alpha,\beta,\gamma \in \mathbb{R}.$$

The corresponding forth order tensor is

$$\mathbb{C} = \alpha\mathbf{I}\otimes\mathbf{I} + \beta\mathbb{I} + \gamma\mathbf{e}_k\otimes\mathbf{I}\otimes\mathbf{e}^k. \tag{A.19}$$

In mechanics, they use various representation theorems for scalar- and tensor-valued functions [6–9]. For example the following theorems hold.

Theorem A.3 *A scalar-valued function f of a symmetric second-order tensor $\mathbf{X} = \mathbf{X}^T$ is a function of three principal invariants of $\mathbf{X}$, that is*

$$f(\mathbf{X}) = f(I_1, I_2, I_3).$$

Theorem A.4 *A scalar-valued function f of an arbitrary second-order tensor is a function of seven invariants of $\mathbf{X}$, that is*

$$f(\mathbf{X}) = f(j_1,\ldots,j_7),$$

where

$$\begin{aligned}
j_1 = \mathrm{tr}\,\mathbf{X},\quad j_2 = \mathrm{tr}\,\mathbf{X}^2,\quad j_3 = \mathrm{tr}\,\mathbf{X}^3,\quad j_4 = \mathrm{tr}(\mathbf{X}\cdot\mathbf{X}^T),\\
j_5 = \mathrm{tr}(\mathbf{X}^2\cdot\mathbf{X}^T),\quad j_6 = \mathrm{tr}(\mathbf{X}^2\cdot\mathbf{X}^{T2}),\quad j_7 = \mathrm{tr}(\mathbf{X}^2\cdot\mathbf{X}^{T2}\cdot\mathbf{X}\cdot\mathbf{X}^T).
\end{aligned} \tag{A.20}$$

Set $j_1, \ldots, j_7$ constitutes the *irreducible polynomial integrity basis* that means that any invariant from the list (A.20) cannot be expressed as a polynomial of other six invariants. These seven invariants are related by only one equation, a so called *syzygy* equation, that however does not define uniquely the seventh invariant through other six invariants, see [7, 10] for details.

The expressions for isotropic, transversely isotropic, orthotropic functions of several vectorial and tensorial arguments are summarized in [7, 9].

Let us present some useful formulae to derive the derivatives of tensor functions. For a real valued function $f(\mathbf{x})$, the first differential is

$$df(\mathbf{x}) = f_{,\mathbf{x}} \cdot d\mathbf{x} = \left. \frac{df(\mathbf{x} + \varepsilon\, d\mathbf{x})}{d\varepsilon} \right|_{\varepsilon=0}.$$

The right-hand side of this equality is termed the *Gâteaux derivative* of f at the point $\mathbf{x}$ in the direction $d\mathbf{x}$. In Cartesian coordinates it is

$$f_{,\mathbf{x}} = \sum_{k=1}^{n} \frac{\partial f}{\partial x_k} \mathbf{i}_k.$$

These definition can be extended to tensorial functions of tensorial arguments in a straightforward manner. For a scalar-valued function $f(\mathbf{X})$, $\mathbf{X} \in \mathbb{T}_2$, the Gâteaux derivative is

$$\left. \frac{d}{d\varepsilon} f(\mathbf{X} + \varepsilon d\mathbf{X}) \right|_{\varepsilon=0} \equiv \lim_{\varepsilon \to 0} \frac{f(\mathbf{X} + \varepsilon d\mathbf{X}) - f(\mathbf{X})}{\varepsilon} = f_{,\mathbf{X}} \cdot\cdot\, d\mathbf{X}^T \tag{A.21}$$

for any tensor $d\mathbf{X}$ that can be non-infinitesimal. In the Cartesian basis this is

$$f_{,\mathbf{X}} = \frac{\partial f}{\partial X_{mn}} \mathbf{i}_m \otimes \mathbf{i}_n,$$

where the X_{mn} are the Cartesian components of $\mathbf{X}$. Then $df = f_{,\mathbf{X}} \cdot\cdot\, d\mathbf{X}^T$. Expression $f_{,\mathbf{X}}$ is called the derivative of f with respect to the tensor argument $\mathbf{X}$. In spite of the fact that the last formula is derived in Cartesian coordinates, it holds in any basis.

For a function that maps $\mathbf{X} \in \mathbb{T}_2$ to $\mathbb{T}_2$, the derivative $\mathbf{F}_{,\mathbf{X}}(\mathbf{X})$ is

$$\mathbf{F}_{,\mathbf{X}} \cdot\cdot\, d\mathbf{X}^T = \left. \frac{\partial}{\partial \varepsilon} \mathbf{F}(\mathbf{X} + \varepsilon d\mathbf{X}) \right|_{\varepsilon=0} \equiv \lim_{\varepsilon \to 0} \frac{\mathbf{F}(\mathbf{X} + \varepsilon d\mathbf{X}) - \mathbf{F}(\mathbf{X})}{\varepsilon}. \tag{A.22}$$

$\mathbf{F}_{,\mathbf{X}}(\mathbf{X})$ is a fourth-order tensor. In the component form it is $\mathbf{F}_{,\mathbf{X}} = \frac{\partial f_{ij}}{\partial x_{mn}} \mathbf{i}_i \otimes \mathbf{i}_j \otimes \mathbf{i}_m \otimes \mathbf{i}_n$. For the derivatives the following notations can be used:

$$f_{,\mathbf{X}}(\mathbf{X}) = \frac{df(\mathbf{X})}{d\mathbf{X}} \quad \text{and} \quad \mathbf{F}_{,\mathbf{X}}(\mathbf{X}) = \frac{d\mathbf{F}(\mathbf{X})}{d\mathbf{X}}.$$

In a similar way, we can define the derivative of tensor-valued functions of any order depending on any tensorial arguments.

Partial derivatives of scalar-valued $f(\mathbf{X}_1, \ldots, \mathbf{X}_m)$ and tensor-valued $\mathbf{F}(\mathbf{X}_1, \ldots, \mathbf{X}_m)$ functions in several tensorial arguments are defined as follows

$$\frac{\partial f}{\partial \mathbf{X}_i} \cdot\cdot \mathbf{Y}^T = \frac{d}{d\varepsilon} f(\mathbf{X}_1, \ldots, \mathbf{X}_i + \varepsilon \mathbf{Y}, \ldots, \mathbf{X}_m)\Big|_{\varepsilon=0},$$

$$\frac{\partial \mathbf{F}}{\partial \mathbf{X}_i} \cdot\cdot \mathbf{Y}^T = \frac{d}{d\varepsilon} \mathbf{F}(\mathbf{X}_1, \ldots, \mathbf{X}_i + \varepsilon \mathbf{y}, \ldots, \mathbf{X}_m)\Big|_{\varepsilon=0}.$$

for any second-order tensor $\mathbf{Y}$. In particular, the following formulae are valid:

$$I_1(\mathbf{X})_{,\mathbf{X}} = \mathbf{I}, \quad I_2(\mathbf{X})_{,\mathbf{X}} = I_1(\mathbf{X})\mathbf{I} - \mathbf{X}^T, I_3(\mathbf{X})_{,\mathbf{X}} = [\mathbf{X}^2 - I_1(\mathbf{X})\mathbf{X} + I_2(\mathbf{X})\mathbf{I}]^T,$$
$$(\operatorname{tr} \mathbf{X}^n)_{,\mathbf{X}} = n\mathbf{X}^T, \quad (\mathbb{C} \cdot\cdot \mathbf{X}^T)_{,\mathbf{X}} = \mathbb{C}, \quad \mathbf{X}_{,\mathbf{X}} = \mathbb{I},$$

where n is an integer number and $\mathbb{C}$ is a fourth-order tensor.

More details on tensorial functions and their properties and applications can be found in [1, 6–10].

A.8 Vector and Tensor Fields

A tensor field is a mapping that assigns a certain order tensor to each point of some domain in $\mathbb{R}^3$. The position of a domain point is determined by three parameters q^1, q^2, q^3 that are curvilinear coordinates in space. Between (q^1, q^2, q^3) and cartesian coordinates (x^1, x^2, x^3) there is one-to-one correspondence except, possibly, some singular points

$$x^i = x^i(q^1, q^2, q^3), \quad q^i = x^i(x^1, x^2, x^3) \quad (i = 1, 2, 3). \tag{A.23}$$

The relations can be written as one vectorial equation

$$\mathbf{r} = \mathbf{r}(q^1, q^2, q^3), \quad \mathbf{r} = x_k \mathbf{i}_k.$$

We will suppose function $\mathbf{r}(q^1, q^2, q^3)$ to be so smooth as we need. As the transformation should be invertible it follows

$$J \equiv \sqrt{g} \equiv \left|\frac{\partial x^i}{\partial q^j}\right| \neq 0 \tag{A.24}$$

anywhere excerpt possible singular points.

Differentiating $\mathbf{r}$ with respect to q^k at each point we get three vectors

$$\mathbf{r}_1 = \frac{\partial \mathbf{r}(q^1, q^2, q^3)}{\partial q^1}, \quad \mathbf{r}_2 = \frac{\partial \mathbf{r}(q^1, q^2, q^3)}{\partial q^2}, \quad \mathbf{r}_3 = \frac{\partial \mathbf{r}(q^1, q^2, q^3)}{\partial q^3}.$$

The vectors are tangent to the corresponding coordinate lines, that is $\mathbf{r}_1$ is tangent to q^1, etc. If (q^1, q^2, q^3) is not a singular point, i.e. $J(q^1, q^2, q^3) \neq 0$, then $\mathbf{r}_i$ are non-coplanar and constitute a basis at a point.

The dual basis to the principal basis is denoted by $\mathbf{r}^i$:

$$\mathbf{r}^i \cdot \mathbf{r}_j = \delta^i_j.$$

The metric coefficients for the coordinate system are

$$g_{ij} = \mathbf{r}_i \cdot \mathbf{r}_j, \qquad g^{ij} = \mathbf{r}^i \cdot \mathbf{r}^j, \qquad g^j_i = \mathbf{r}_i \cdot \mathbf{r}^j = \delta^j_i.$$

Through the metric coefficients we can express the squared length of infinitesimal vector $\mathbf{dr}$

$$(\mathrm{d}s)^2 = \mathbf{dr} \cdot \mathbf{dr} = \mathbf{r}_i \mathrm{d}q^i \cdot \mathbf{r}_j \mathrm{d}q^j = g_{ij} \mathrm{d}q^i \mathrm{d}q^j, \quad \mathbf{dr} = \frac{\partial \mathbf{r}}{\partial q^i} \mathrm{d}q^i = \mathbf{r}_i \mathrm{d}q^i. \qquad \text{(A.25)}$$

Let $f(q^1, q^2, q^3)$ be a scalar function of q^i. Its first differential is

$$\mathrm{d}f(q^1, q^2, q^3) = \frac{\partial f(q^1, q^2, q^3)}{\partial q^i} \mathrm{d}q^i = \mathbf{r}^i \frac{\partial f}{\partial q^i} \cdot \mathbf{dr}.$$

Let us introduce a formal vector

$$\nabla = \mathbf{r}^i \frac{\partial}{\partial q^i}$$

called Hamilton's *nabla-operator*. It acts on function f by the rule

$$\nabla f = \mathbf{r}^i \frac{\partial f}{\partial q^i}.$$

Nabla-operator is also called the gradient.

Using ∇ we represent the first differential of $\mathbf{f}(q^1, q^2, q^3)$ as follows

$$\mathbf{df} = \mathbf{dr} \cdot \nabla \mathbf{f} = \nabla \mathbf{f}^T \cdot \mathbf{dr}.$$

Quantity $\nabla \mathbf{f}$ is called the gradient of $\mathbf{f}$, it is a second-order tensor. Note that sometimes they call $\nabla \mathbf{f}^T$ by the gradient of $\mathbf{f}$. The difference in the definition of the gradient as well of the divergence and the rotor can bring uncertainness. We will use the notation $\mathbf{f}$ for $\nabla \mathbf{f}^T$. $\nabla \mathbf{f}^T$ is also called the derivative of $\mathbf{f}$ with respect to $\mathbf{r}$, it is denoted by

$$\frac{\mathbf{df}}{\mathbf{dr}} = \nabla \mathbf{f}^T = \operatorname{grad} \mathbf{f}.$$

Considering ∇ as a formal vector, we can also introduce its dot and cross products with vector $\mathbf{f} = \mathbf{f}(q^1, q^2, q^3)$ that brings us the definition of the *divergence* and *rotor* (*curl*) of vector field $\mathbf{f}$:

$$\nabla \cdot \mathbf{f} = \mathbf{r}^i \cdot \frac{\partial \mathbf{f}}{\partial q^i}, \qquad \operatorname{div} \mathbf{f} = \frac{\partial \mathbf{f}}{\partial q^i} \cdot \mathbf{r}^i = \nabla \cdot \mathbf{f},$$

$$\nabla \times \mathbf{f} = \mathbf{r}^i \times \frac{\partial \mathbf{f}}{\partial q^i}, \qquad \operatorname{rot} \mathbf{f} = \frac{\partial \mathbf{f}}{\partial q^i} \times \mathbf{r}^i = -\nabla \times \mathbf{f}.$$

By similar formulae we introduce the divergence and rotor for a tensor field $\mathbf{A}$ of any order

$$\nabla\cdot\mathbf{A} = \mathbf{r}^i\cdot\frac{\partial}{\partial q^i}\mathbf{A}, \qquad \operatorname{div}\mathbf{A} = \frac{\partial\mathbf{A}}{\partial q^i}\cdot\mathbf{r}^i \neq \nabla\cdot\mathbf{A},$$

$$\nabla\times\mathbf{A} = \mathbf{r}^i\times\frac{\partial}{\partial q^i}\mathbf{A}, \qquad \operatorname{rot}\mathbf{A} = \frac{\partial\mathbf{A}}{\partial q^i}\times\mathbf{r}^i.$$

Differentiating vector and tensor functions we should know the derivatives of the basis vector as the derivative of a vector function is

$$\begin{aligned}\frac{\partial}{\partial q^k}\mathbf{f}(q^1,q^2,q^3) &= \frac{\partial}{\partial q^k}\left[f^i(q^1,q^2,q^3)\mathbf{r}_i\right]\\ &= \frac{\partial f^i(q^1,q^2,q^3)}{\partial q^k}\mathbf{r}_i + f^i(q^1,q^2,q^3)\frac{\partial\mathbf{r}_i}{\partial q^k}.\end{aligned}$$

Note the derivatives of basis vectors have the following symmetry property

$$\frac{\partial}{\partial q^j}\mathbf{r}_i = \frac{\partial}{\partial q^j}\left(\frac{\partial}{\partial q^i}\mathbf{r}\right) = \frac{\partial^2\mathbf{r}}{\partial q^j\partial q^i} = \frac{\partial^2\mathbf{r}}{\partial q^i\partial q^j} = \frac{\partial}{\partial q^i}\mathbf{r}_j.$$

As $\mathbf{r}_i(i = 1,2,3)$ is a basis, we can expand $\frac{\partial}{\partial q^j}\mathbf{r}_i$

$$\frac{\partial}{\partial q^j}\mathbf{r}_i = \Gamma^k_{ij}\mathbf{r}_k, \tag{A.27}$$

where Γ^k_{ij} are called the *Christoffel symbols* of second kind. By (A.26), Christoffel symbols also possess symmetry properties

$$\Gamma^k_{ij} = \Gamma^k_{ji}. \tag{A.28}$$

Now we have

$$\frac{\partial}{\partial q^k}\mathbf{f}(q^1,q^2,q^3) = \left(\frac{\partial f^i}{\partial q^k} + \Gamma^i_{kt}f^t\right)\mathbf{r}_i. \tag{A.29}$$

This operation is called the covariant differentiation. The coefficient at $\mathbf{r}_i$ is called the covariant derivative of vector function $\mathbf{f}$, it is denoted by

$$\nabla_k f^i = \frac{\partial f^i}{\partial q^k} + \Gamma^i_{kt}f^t.$$

Christoffel symbols can be expressed through the metric coefficients

$$\Gamma^k_{ij} = \frac{1}{2}g^{kt}\left(\frac{\partial g_{it}}{\partial q^j} + \frac{\partial g_{tj}}{\partial q^i} - \frac{\partial g_{ji}}{\partial q^t}\right). \tag{A.30}$$

Along with the second kind Christoffel symbols, we introduce *Christoffel symbols of the first kind*, they are denoted by Γ_{ijk}:

$$\Gamma_{ijt} = \Gamma^k_{ij} g_{kt} = \frac{1}{2}\left(\frac{\partial g_{it}}{\partial q^j} + \frac{\partial g_{tj}}{\partial q^i} - \frac{\partial g_{ji}}{\partial q^t}\right). \tag{A.31}$$

The derivatives of the dual basis are also expressed with use of Christoffel symbols:

$$\frac{\partial \mathbf{r}^j}{\partial q^i} = -\Gamma^j_{it}\mathbf{r}^t. \tag{A.32}$$

The differentiation formulae of a vector field take the form

$$\frac{\partial \mathbf{f}}{\partial q^i} = \mathbf{r}^k \nabla_i f_k = \mathbf{r}_j \nabla_i f^j,$$

where

$$\nabla_k f_i = \frac{\partial f_i}{\partial q^k} - \Gamma^j_{ki} f_j, \qquad \nabla_k f^i = \frac{\partial f^i}{\partial q^k} + \Gamma^i_{kt} f^t.$$

For the vector gradient we get

$$\nabla \mathbf{f} = \mathbf{r}^i \otimes \mathbf{r}^j \nabla_i f_j = \mathbf{r}^i \otimes \mathbf{r}_j \nabla_i f^j$$

that shows that $\nabla_i f_j$ are covariant components of $\nabla \mathbf{f}$ whereas $\nabla_i f^j$ its mixed components.

The differentiation formulae for a second-order tensor field $\mathbf{A}$ are awkward as we should differentiate the dyads of basis vectors. For example

$$\frac{\partial}{\partial q^k}\mathbf{A} = \nabla_k a^{ij}\mathbf{r}_i \otimes \mathbf{r}_j = \nabla_k a_{ij}\mathbf{r}^i \otimes \mathbf{r}^j,$$

where

$$\nabla_k a^{ij} = \frac{\partial a^{ij}}{\partial q^k} + \Gamma^i_{ks} a^{sj} + \Gamma^j_{ks} a^{is}, \qquad \nabla_k a_{ij} = \frac{\partial a_{ij}}{\partial q^k} - \Gamma^s_{ki} a_{sj} - \Gamma^s_{kj} a_{is}.$$

Using the identity

$$\Gamma^i_{in} = \frac{1}{\sqrt{g}}\frac{\partial \sqrt{g}}{\partial q^n},$$

we can write out the divergence and rotor of vector and tensor fields in the form that does not contain Christoffel symbols:

$$\nabla\cdot\mathbf{f} = \frac{1}{\sqrt{g}}\frac{\partial}{\partial q^i}\left(\sqrt{g} f^i\right), \quad \nabla\cdot\mathbf{A} = \frac{1}{\sqrt{g}}\frac{\partial}{\partial q^i}\left(\sqrt{g} a^{ij}\mathbf{r}_j\right). \tag{A.33}$$

Curvilinear coordinates (q^1, q^2, q^3) are widely used in various integral formulae. Let $f(x_1, x_2, x_3)$ be a continuous function of Cartesian coordinates (x_1, x_2, x_3) in some domain V. Changing the Cartesian coordinates to (q^1, q^2, q^3) we present f in the form $f = f(q^1, q^2, q^3)$.

The integral of f over V is

$$\iiint_V f(x_1, x_2, x_3)\mathrm{d}x_1\mathrm{d}x_2\mathrm{d}x_3 = \iiint_V f(q^1, q^2, q^3)J\mathrm{d}q^1\mathrm{d}q^2\mathrm{d}q^3, \quad J = \sqrt{g} = \left|\frac{\partial x_i}{\partial q^j}\right|.$$

We remind the *Gauss–Ostrogradsky theorem* that consists of the formula

$$\iiint_V \nabla f \mathrm{d}V = \iint_S f\mathbf{n}\,\mathrm{d}S$$

for a scalar function f. For a vector function $\mathbf{f}$ and tensor function $\mathbf{A}$, the Gauss–Ostrogradsky formulae are

$$\iiint_V \nabla\mathbf{f}\,\mathrm{d}V = \iint_S \mathbf{n}\otimes\mathbf{f}\,\mathrm{d}S, \quad \iiint_V \nabla\mathbf{A}\,\mathrm{d}V = \iint_S \mathbf{n}\otimes\mathbf{A}\,\mathrm{d}S.$$

From these, the *divergence theorems* follow

$$\iiint_V \nabla\cdot\mathbf{f}\,\mathrm{d}V = \iint_S \mathbf{n}\cdot\mathbf{f}\,\mathrm{dS}, \qquad \iiint_V \mathrm{div}\,\mathbf{f}\,\mathrm{d}V = \iint_S \mathbf{f}\cdot\mathbf{n}\,\mathrm{dS}, \tag{A.34}$$

$$\iiint_V \nabla\cdot\mathbf{a}\,\mathrm{d}V = \iint_S \mathbf{n}\cdot\mathbf{A}\,\mathrm{d}S, \qquad \iiint_V \mathrm{div}\,\mathbf{a}\,\mathrm{d}V = \iint_S \mathbf{A}\cdot\mathbf{n}\,\mathrm{d}S. \tag{A.35}$$

There are other useful integral formulae, for example

$$\iiint_V \nabla\times\mathbf{f}\,\mathrm{d}V = \iint_S \mathbf{n}\times\mathbf{f}\,\mathrm{d}S, \quad \iiint_V \nabla\mathbf{f}^T\,\mathrm{d}V = \iint_S \mathbf{f}\otimes\mathbf{n}\,\mathrm{d}S,$$

$$\iiint_V \nabla\times\mathbf{A}\,\mathrm{d}V = \iint_S \mathbf{n}\times\mathbf{A}\,\mathrm{d}S.$$

The above formulae relate the volume integrals with the ones over the volume surfaces. There exist the formulae relating the surface integrals with the integrals over the surface contours, they are based on the Stokes formula. Let vector function $\mathbf{f}$ be given on one-connected surface S with boundary contour Γ. *Stokes's formula* is

$$\oint_\Gamma \mathbf{f}\cdot\mathbf{dr} = \iint_S (\mathbf{n}\times\nabla)\cdot\mathbf{f}\,\mathrm{d}S.$$

For a second-order tensor $\mathbf{A}$, Stokes's formula takes the form

$$\oint_\Gamma \mathbf{dr}\cdot\mathbf{a} = \iint_S (\mathbf{n}\times\nabla)\cdot\mathbf{A}\,\mathrm{d}S \quad \text{or} \quad \oint_\Gamma \mathbf{A}\cdot\mathbf{dr} = \iint_S (\mathbf{n}\times\nabla)\cdot\mathbf{A}^T\,\mathrm{d}S.$$

Let us note that (A.36) is valid for tensors of any order.

A.9 Curves in Space

The equation of a spatial curve ℓ is

$$\mathbf{r} = \mathbf{r}(t).$$

In Cartesian coordinates it is

$$x_1 = x_1(t), \quad x_2 = x_2(t), \quad x_3 = x_3(t),$$

$t \in \mathbb{R}$ denotes a parameter. We suppose $\mathbf{r}(t)$ to be doubly differentiable with respect to t.

The length of the part of curve ℓ corresponding to $t \in [t_1, t_2]$ is

$$s(t_1, t_2) = \int_{t_1}^{t_2} \sqrt{\dot{x}_1^2 + \dot{x}_2^2 + \dot{x}_3^2}\mathrm{d}t \equiv \int_{t_1}^{t_2} |\dot{\mathbf{r}}(t)|\mathrm{d}t, \qquad (\dot{\ldots}) = \frac{\mathrm{d}}{\mathrm{d}t}(\ldots).$$

Instead of parameter t we can use the length parameter s. The s-parametrization of ℓ is called *natural*, $\mathbf{r} = \mathbf{r}(s)$.

At a point, the unit vector tangent to ℓ is

$$\boldsymbol{\tau} = \mathbf{r}'(s), \qquad (\ldots)' = \frac{\mathrm{d}}{\mathrm{d}s}(\ldots).$$

Differentiating $\boldsymbol{\tau}$ we get $\boldsymbol{\tau}' = k\boldsymbol{\nu}$, where $\boldsymbol{\nu}$ is the principal unit normal to ℓ at the point and k is its *curvature*.

As $\boldsymbol{\tau}$ is a unit vector, $\boldsymbol{\nu}$ is orthogonal to $\boldsymbol{\tau}$. Vector $\boldsymbol{\beta} = \boldsymbol{\tau} \times \boldsymbol{\nu}$ is unit and orthogonal to $\boldsymbol{\tau}$ and $\boldsymbol{\nu}$, it is called the *binormal*. Vectors $\boldsymbol{\tau}$, $\boldsymbol{\nu}$, and $\boldsymbol{\beta}$ constitute the so called *moving trihedron* to ℓ or the natural basis on ℓ (Fig. A.2). For the basis vectors, *Frenet–Serret formulae* hold

$$\boldsymbol{\tau}' = k\boldsymbol{\nu}, \quad \boldsymbol{\nu}' = -k\boldsymbol{\tau} - \varkappa\boldsymbol{\beta}, \quad \boldsymbol{\beta}' = \varkappa\boldsymbol{\nu}. \tag{A.37}$$

Coefficient $\varkappa$ is called the curve *torsion*. For a planar curve $\varkappa = 0$. If $k(s)$ and $\varkappa(s)$ are given functions of s the curve is uniquely defined in space up to its translation and rotation. Formulae (A.37) can be presented in a more compact form:

$$\boldsymbol{\tau}' = \boldsymbol{\omega} \times \boldsymbol{\tau}, \quad \boldsymbol{\nu}' = \boldsymbol{\omega} \times \boldsymbol{\nu}, \quad \boldsymbol{\beta}' = \boldsymbol{\omega} \times \boldsymbol{\beta}, \tag{A.38}$$

where $\boldsymbol{\omega}$ is the *Darboux vector*, $\boldsymbol{\omega} = -\varkappa\boldsymbol{\tau} + k\boldsymbol{\beta}$.

Formulae (A.37) or (A.38) are used when we should find the derivatives of vector/tensor fields given on a curve. As an example, we calculate the derivative of a vector field $\mathbf{u}$ given on ℓ. Let

$$\mathbf{u} = u_1\boldsymbol{\tau} + u_2\boldsymbol{\nu} + u_3\boldsymbol{\beta}.$$

Its derivative with respect to s is

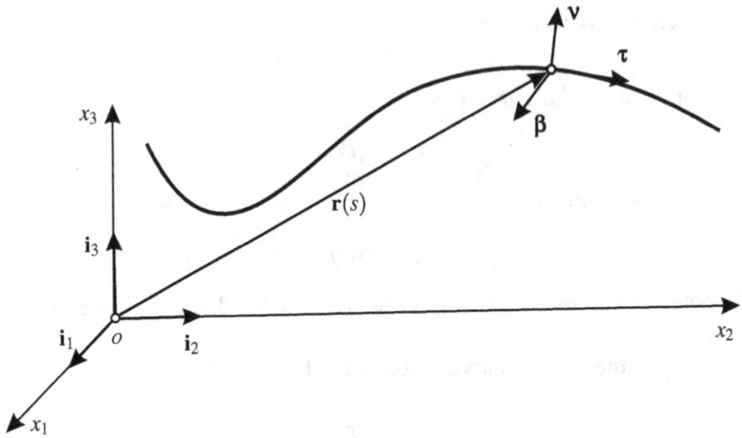

Fig. A.2 Curve in space and its moving trihedron

$$\begin{aligned}\mathbf{u}' &= u_1'\boldsymbol{\tau} + u_2'\boldsymbol{\nu} + u_3'\boldsymbol{\beta} + \boldsymbol{\omega} \times \mathbf{u} \\ &= (u_1' - ku_2)\boldsymbol{\tau} + (u_2' + ku_1 + \varkappa u_3)\boldsymbol{\nu} + (u_3' - \varkappa u_2)\boldsymbol{\beta}.\end{aligned} \tag{A.39}$$

The derivative of a tensor field of any order given in basis $\boldsymbol{\tau}, \boldsymbol{\nu}, \boldsymbol{\beta}$ on ℓ is taken in a similar way.

A.10 Surfaces

In terms of the position vector, the equation of surface Σ is

$$\boldsymbol{\rho} = \boldsymbol{\rho}(q^1, q^2),$$

where q^1, q^2 are curvilinear coordinates on Σ, $(q^1, q^2) \in \boldsymbol{\Omega} \subset \mathbb{R}^2$. To avoid ambiguity with the results of previous sections, we change $\mathbf{r}$ to another notation: $\boldsymbol{\rho}$. The vectorial equation for Σ is equivalent to three scalar equations

$$x_1 = x_1(q^1, q^2), \quad x_2 = x_2(q^1, q^2), \quad x_3 = x_3(q^1, q^2).$$

As for the curve, we assume $\boldsymbol{\rho}$ to be doubly differentiable function of its arguments.

At a point of Σ, the derivatives of $\boldsymbol{\rho}$ with respect to q^1, q^2 constitute a basis in the tangent plane to Σ

$$\boldsymbol{\rho}_1 = \frac{\partial \boldsymbol{\rho}}{\partial q^1}, \quad \boldsymbol{\rho}_2 = \frac{\partial \boldsymbol{\rho}}{\partial q^2}.$$

Vectors $\boldsymbol{\rho}_1$ and $\boldsymbol{\rho}_2$ are tangent to the corresponding coordinate lines on Σ, see Fig. A.3.

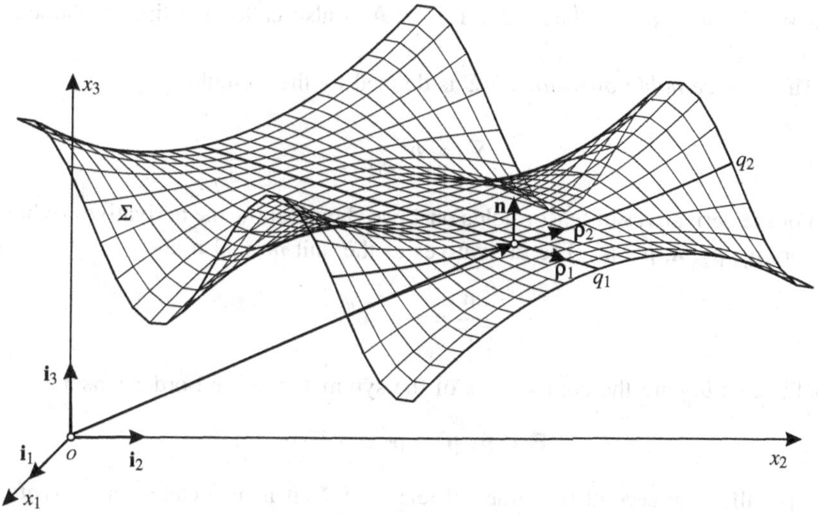

Fig. A.3 Surface Σ and the coordinate lines for (q^1, q^2)

At a point, the unit normal vector $\mathbf{n}$ to Σ is

$$\mathbf{n} = \frac{\boldsymbol{\rho}_1 \times \boldsymbol{\rho}_2}{|\boldsymbol{\rho}_1 \times \boldsymbol{\rho}_2|}.$$

At a point of Σ, vectors $\boldsymbol{\rho}_1$, $\boldsymbol{\rho}_2$, $\mathbf{n}$ constitute a basis in $\mathbb{R}^3$. The dual basis is $(\boldsymbol{\rho}^1, \boldsymbol{\rho}^2, \mathbf{n})$, where

$$\boldsymbol{\rho}_\alpha \cdot \boldsymbol{\rho}^\beta = \delta_\alpha^\beta \quad (\alpha, \beta = 1, 2).$$

Finding the length of a curve on the surface or the area of some part of Σ are important for applications. We start with the length of an elementary segment on Σ that is

$$ds^2 = d\boldsymbol{\rho} \cdot d\boldsymbol{\rho} = \boldsymbol{\rho}_i dq^i \cdot \boldsymbol{\rho}_j dq^j = a_{ij} dq^i dq^j,$$

where $a_{\alpha\beta} = \boldsymbol{\rho}_\alpha \cdot \boldsymbol{\rho}_\beta$ are *metric coefficients* of Σ. Form $a_{ij} dq^i dq^j$ is called the first quadratic form of Σ. Similarly we introduce the metric coefficients $a^{\alpha\beta} = \boldsymbol{\rho}^\alpha \cdot \boldsymbol{\rho}^\beta$. The area of an infinitesimal parallelogram with sides $\boldsymbol{\rho}_k dq^k$ is

$$da = |\boldsymbol{\rho}_1 \times \boldsymbol{\rho}_2| dq^1 dq^2.$$

Let us introduce the metric tensor $\mathbf{A}$:

$$\mathbf{A} = \boldsymbol{\rho}^\alpha \otimes \boldsymbol{\rho}_\alpha = \mathbf{i} - \mathbf{n} \otimes \mathbf{n}.$$

Tensor $\mathbf{A}$ plays the role of the unit tensor acting in the tangent plane to Σ. If $\mathbf{u}$ is in the tangent plane to Σ at a point, that is $\mathbf{u} \cdot \mathbf{n} = 0$, then $\mathbf{A} \cdot \mathbf{u} = \mathbf{u}$. It can be shown that $\mathbf{A} = a_{\alpha\beta}\boldsymbol{\rho}^\alpha \otimes \boldsymbol{\rho}^\beta = a^{\alpha\beta}\boldsymbol{\rho}_\alpha \otimes \boldsymbol{\rho}_\beta$. Tensor $\mathbf{A}$ is also called the first fundamental tensor of Σ.

The *surface nabla-operator* on Σ is defined by the formula

$$\nabla_s = \boldsymbol{\rho}^\alpha \frac{\partial}{\partial q^\alpha}.$$

For the derivatives of tensor fields on Σ we should know the derivatives of basis vectors $\boldsymbol{\rho}_1$, $\boldsymbol{\rho}_2$, $\mathbf{n}$, $\boldsymbol{\rho}^1$, $\boldsymbol{\rho}^2$. The derivatives of the unit normal $\mathbf{n}$ are

$$\frac{\partial \mathbf{n}}{\partial q^\alpha} = -b_{\alpha\beta}\boldsymbol{\rho}^\beta.$$

Coefficients $b_{\alpha\beta}$ are the components of the symmetric second-order tensor

$$\mathbf{B} = b_{\alpha\beta}\boldsymbol{\rho}^\alpha \otimes \boldsymbol{\rho}^\beta \equiv -\nabla_s \mathbf{n}$$

that is called the second fundamental tensor of Σ, it is also called the *curvature tensor*. The components of $\mathbf{B}$ are the coefficients of the second quadratic form of Σ. As any symmetric tensor, $\mathbf{B}$ has real eigenvalues denoted by k_1 and k_2 and corresponding eigenvectors $\mathbf{e}_1$ and $\mathbf{e}_2$. Eigenvalues k_1 and k_2 are called *principal curvatures* of Σ at a point. The lines on Σ, for which the tangent vectors are $\mathbf{e}_1$ and $\mathbf{e}_2$, are called the *principal curvature lines*.

The *mean curvature* H and *Gaussian curvature* K are important characteristics of Σ:

$$H = \frac{1}{2}\operatorname{tr}\mathbf{B} = \frac{1}{2}(k_1 + k_2), \quad K = I_2(\mathbf{B}) = k_1 k_2.$$

The derivatives of $\boldsymbol{\rho}_\alpha$, and $\boldsymbol{\rho}^\alpha$ are

$$\frac{\partial \boldsymbol{\rho}_\alpha}{\partial q^\beta} = \Gamma^\gamma_{\alpha\beta}\boldsymbol{\rho}_\gamma + b_{\alpha\beta}\mathbf{n}, \quad \frac{\partial \boldsymbol{\rho}^\alpha}{\partial q^\beta} = -\Gamma^\alpha_{\beta\gamma}\boldsymbol{\rho}^\gamma + b^\alpha_\beta \mathbf{n}.$$

We remind that $\Gamma^\gamma_{\alpha\beta}$ are Christoffel symbols relating with the metric coefficients by (A.30).

We will use some formulae for the gradient of vector/tensor fields given on Σ. Let us introduce

$$\mathbf{u} = \mathbf{u}_s + w\mathbf{n}, \quad \mathbf{u}_s = u_1(q^1, q^2)\boldsymbol{\rho}^1 + u_2(q^1, q^2)\boldsymbol{\rho}^2, \quad w = u_3 = u^3 = \mathbf{u}\cdot\mathbf{n},$$

$$\mathbf{T} = T^{\alpha\beta}\boldsymbol{\rho}_\alpha \otimes \boldsymbol{\rho}_\beta + T^{3\beta}\mathbf{n} \otimes \boldsymbol{\rho}_\beta + T^{\alpha 3}\boldsymbol{\rho}_\alpha \otimes \mathbf{n} + T^{33}\mathbf{n} \otimes \mathbf{n}.$$

Then

$$\nabla_s \mathbf{u} = (\nabla_s \mathbf{u}_s)\cdot\mathbf{A} - w\mathbf{B} + (\nabla_s w + \mathbf{B}\cdot\mathbf{u}_s) \otimes \mathbf{n}, \tag{A.40}$$

$$\begin{aligned}\nabla_s \cdot \mathbf{T} &= \boldsymbol{\rho}^\gamma \frac{\partial}{\partial q^\gamma} \cdot \left(T^{\alpha\beta} \boldsymbol{\rho}_\alpha \otimes \boldsymbol{\rho}_\beta + T^{3\beta} \mathbf{n} \otimes \boldsymbol{\rho}_\beta + T^{\alpha 3} \boldsymbol{\rho}_\alpha \otimes \mathbf{n} + T^{33} \mathbf{n} \otimes \mathbf{n}\right) \\ &= \frac{\partial T^{\alpha\beta}}{\partial q^\alpha} \boldsymbol{\rho}_\beta + T^{\alpha\beta} \Gamma^\gamma_{\alpha\gamma} \boldsymbol{\rho}_\beta + T^{\alpha\beta} \Gamma^\gamma_{\beta\alpha} \boldsymbol{\rho}_\gamma + T^{\alpha\beta} b_{\alpha\beta} \mathbf{n} - T^{3\beta} b^\gamma_\gamma \boldsymbol{\rho}_\beta \\ &\quad - T^{33} b^\gamma_\gamma \mathbf{n} + \frac{\partial T^{\alpha 3}}{\partial q^\alpha} \mathbf{n} + T^{\alpha 3} \Gamma^\gamma_{\alpha\gamma} \mathbf{n} - T^{\alpha 3} b^\beta_\alpha \boldsymbol{\rho}_\beta. \end{aligned} \tag{A.41}$$

Using Stokes's formula (A.36), we can establish the *divergence theorem* on the surface, that is

$$\iint_\Sigma (\nabla_s \cdot \mathbf{T} + 2H \mathbf{n} \cdot \mathbf{T}) \mathrm{d}\Sigma = \int_{\partial\Sigma} \boldsymbol{\nu} \cdot \mathbf{T} \, \mathrm{d}s, \tag{A.42}$$

where $\boldsymbol{\nu}$ is the unit external normal to contour $\partial\Sigma$, lying in the tangent plane to Σ, that is $\boldsymbol{\nu} \cdot \mathbf{n} = 0$. Using (A.42) we can derive the following useful integral formulae

$$\iint_\Sigma (\nabla_s \mathbf{T} + 2H \mathbf{n} \otimes \mathbf{T}) \mathrm{d}\Sigma = \int_{\partial\Sigma} \boldsymbol{\nu} \otimes \mathbf{T} \, \mathrm{d}s,$$

$$\iint_\Sigma (\nabla_s \times \mathbf{T} + 2H \mathbf{n} \times \mathbf{T}) \mathrm{d}\Sigma = \int_{\partial\Sigma} \boldsymbol{\nu} \times \mathbf{T} \, \mathrm{d}s.$$

Note that (A.42) and (A.43) are valid for tensor $\mathbf{T}$ of any order.

Appendix B
Elements of Rigid Body Dynamics

Dynamics of rigid bodies is one of the oldest parts of mechanics. Such notions of rigid body mechanics as the moment, the inertia tensor and others are also used in micropolar mechanics. We recall some notions and equations of classical mechanics that are used throughout the book in brief. For details we can refer to various textbooks one of which is [11].

The principle notion of classical mechanics is the *mass-point* or the *material point* or the *particle*. A mass-point possesses a mass m, its position is given by a radius vector $\mathbf{r}(t)$ at instant time t. The vector $\mathbf{r}(t)$ connects some given origin O, called also pole, with the point. In an inertial coordinate system, the motion of the material point is determined by *Newton's law*

$$m\dot{\mathbf{v}} = \mathbf{f}, \tag{B.1}$$

where $\mathbf{v} = \dot{\mathbf{r}}$ is the velocity, $\mathbf{f}$ is the force acting on the mass-point, and the overdot denotes the derivative with respect to t, for example, $\dot{\mathbf{r}} = \mathrm{d}\mathbf{r}/\mathrm{d}t$.

Definition B.1 The momentum of the mass-point called also the linear momentum, is the quantity $\mathfrak{P} = m\mathbf{v}$. The moment of momentum of the mass-point called also the angular momentum, with respect to a point o with radius vector $\mathbf{r}_0$ is $\mathfrak{M} = (\mathbf{r} - \mathbf{r}_0) \times m\mathbf{v}$.

The motion of n mass-points is determined by n vectorial equations

$$m_i\dot{\mathbf{v}}_i = \mathbf{f}_i, \qquad i = 1, 2, \ldots, n, \tag{B.2}$$

where $\mathbf{v}_i = \dot{\mathbf{r}}_i$, m_i and $\mathbf{r}_i(t)$ are the mass and the position vector of the ith point, respectively, $\mathbf{f}_i$ is the force acting on the ith mass-point.

Definition B.2 The momentum and the moment of momentum of n mass-points with respect to the point o with the radius-vector $\mathbf{r}_0$ are

V. A. Eremeyev et al., *Foundations of Micropolar Mechanics*,
SpringerBriefs in Continuum Mechanics,
DOI: 10.1007/978-3-642-28353-6,

$$\mathfrak{P} = \sum_{i=1}^{n} m_i \mathbf{v}_i \quad \text{and} \quad \mathfrak{M} = \sum_{i=1}^{n} (\mathbf{r}_i - \mathbf{r}_0) \times m_i \mathbf{v}_i,$$

respectively.

For the motion of the mass-point system the following results hold.

Theorem B.1 *The rate of the change of the momentum of n mass-points is equal to the total (resultant) force vector $\mathfrak{F}$, that is the sum of all the forces acting to the mass-points*

$$\frac{\mathrm{d}}{\mathrm{d}t}\mathfrak{P} = \mathfrak{F}, \qquad \mathfrak{F} \triangleq \sum_{i=1}^{n} \mathbf{f}_i. \tag{B.3}$$

Theorem B.2 *The rate of the change of the moment of momentum with respect to pole $\mathbf{r}_0$ of n mass-points is equal to the total torque (resultant moment) $\mathfrak{C}$ with respect to point o of all forces acting on the mass-points*

$$\frac{\mathrm{d}}{\mathrm{d}t}\mathfrak{M} = \mathfrak{C}, \qquad \mathfrak{C} \triangleq \sum_{i=1}^{n} (\mathbf{r}_i - \mathbf{r}_0) \times \mathbf{f}_i. \tag{B.4}$$

In the case of n mass-points the balance equations of the momentum (B.3) and the moment of momentum (B.4) are consequences of the motion equations (B.2).

The *kinetic energy* of the n mass-points is

$$K \triangleq \frac{1}{2} \sum_{i=1}^{n} m_i \mathbf{v}_i \cdot \mathbf{v}_i.$$

A more complex mechanical object is the *rigid body.*

Definition B.3 A set of material points for which the mutual distances between the points remain unchanged in motion, is called the rigid body.

The rigid body position in space is uniquely determined by the position of one of its points and the body orientation in the space. Indeed, let $o \in \mathscr{P}$ be a point the body called the pole and $\mathbf{r}_0(t)$ is the position vector. Let us rigidly "imbed" the coordinate trihedron with unit vectors $\mathbf{d}_1(t)$, $\mathbf{d}_2(t)$, $\mathbf{d}_3(t)$, $\mathbf{d}_i \cdot \mathbf{d}_j = \delta_{ij}$ into the body, see Fig. B.1. Then the position of any point $z \in \mathscr{P}$ is given by

$$\mathbf{r}(t) = \mathbf{r}_0(t) + \mathbf{z}(t), \quad \mathbf{z} = z_i \mathbf{d}_i(t). \tag{B.5}$$

For the body, we fix an initial configuration $\varkappa$. For example, we can take the body position at instant $t = 0$ as the initial configuration. The position of the pole o, the point z and the imbedded trihedron of the coordinate axes that are $\mathbf{R}_0$, $\mathbf{R} = \mathbf{R}_0 + \mathbf{Z}$, $\mathbf{D}_1$, $\mathbf{D}_3$, $\mathbf{D}_3$, respectively, in the initial configuration, define the body position uniquely at any instant. As the body is rigid, $\mathbf{Z} = z_i \mathbf{D}_i$.

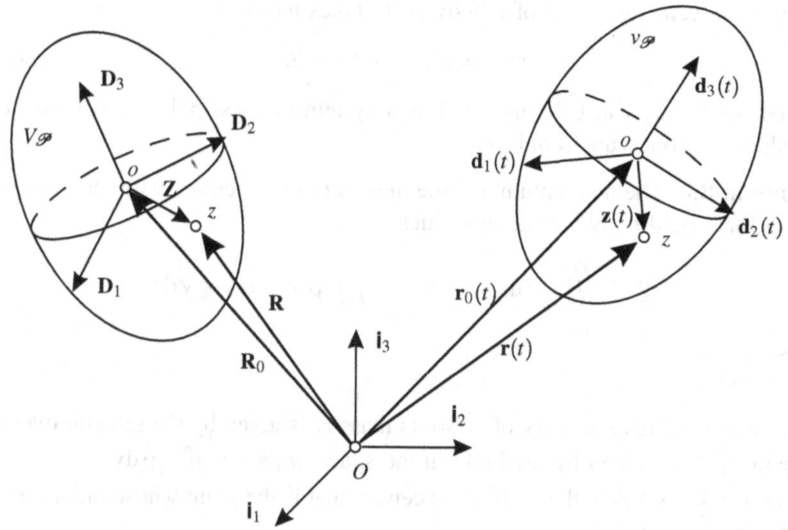

Fig. B.1 Rigid body motion

To describe the body rotation, instead of vectors $\mathbf{d}_i$ we can introduce a proper orthogonal tensor $\mathbf{Q} = \mathbf{d}_i \otimes \mathbf{D}_i$. Then Eq. (B.5) takes the form

$$\mathbf{r}(t) = \mathbf{R}_0 + \mathbf{u}(t) + \mathbf{Q}(t)\cdot\mathbf{Z}. \tag{B.6}$$

Hence the rigid body motion is determined by two quantities, one of which is the translation vector of point o, i.e. $\mathbf{u}(t) = \mathbf{r}_0(t) - \mathbf{R}_0$, and another is the rotation tensor $\mathbf{Q}(t)$. To describe the motion, we also can use *Rodrigues's finite rotation vector* $\boldsymbol{\theta}$, cf. [11] that we see in the representation of the proper orthogonal tensor (A.8). In classical mechanics, to determine the rigid body orientation, one can use other sets of three parameters, for example Euler's angles, airplane or ship angles, etc., see [11]. We should underline that the position of a rigid body $\mathscr{P}$ in the space is always determined by six scalar quantities.

Differentiating (B.6) we get

$$\dot{\mathbf{r}}(t) = \dot{\mathbf{u}}(t) + \dot{\mathbf{Q}}(t)\cdot\mathbf{Z}. \tag{B.7}$$

$\mathbf{Q}$ is orthogonal so tensor $\dot{\mathbf{Q}}\cdot\mathbf{Q}^T$ is skew-symmetric and by (A.5), it can be represented in the form

$$\dot{\mathbf{Q}}\cdot\mathbf{Q}^T = \boldsymbol{\omega} \times \mathbf{I}, \tag{B.8}$$

where $\boldsymbol{\omega}$ is the angular velocity of $\mathscr{P}$. By (A.6), vector $\boldsymbol{\omega}$ can be determined as follows

$$\boldsymbol{\omega} = -\frac{1}{2}(\dot{\mathbf{Q}}\cdot\mathbf{Q}^T)_\times. \tag{B.9}$$

Thus the velocity vector of a body point takes the form

$$\mathbf{v}(t) = \dot{\mathbf{u}}(t) + \boldsymbol{\omega}(t) \times \mathbf{Z}, \tag{B.10}$$

The rigid body can be considered as a system of mass-points and so we can introduce the following definitions.

Definition B.4 The momentum and the moment of momentum with respect to the pole o for a rigid body are the quantities

$$\mathfrak{P} = \iiint\limits_{v_{\mathscr{P}}} \rho \mathbf{v}\, dv, \qquad \mathfrak{M} = \iiint\limits_{v_{\mathscr{P}}} \rho(\mathbf{r} - \mathbf{r}_0) \times \mathbf{v}\, dv,$$

respectively.

Here ρ is the mass density of $\mathscr{P}$ so its mass m is given by the integral over the domain $v_{\mathscr{P}} \subset \mathbb{R}^3$ taken by the body in the space, $m(\mathscr{P}) = \iiint_{v_{\mathscr{P}}} \rho\, \mathbf{d}v$.

Let us take as a pole the body mass center, that is the point whose radius vector $\mathbf{r}_0$ satisfies the relation

$$\iiint\limits_{v_{\mathscr{P}}} \rho(\mathbf{r} - \mathbf{r}_0) dv = \mathbf{0}.$$

Then the momentum and the moment of momentum of the rigid body take the form

$$\mathfrak{P} = m\mathbf{v}_0, \qquad \mathfrak{M} = \iiint\limits_{v_{\mathscr{P}}} \rho \mathbf{z} \times \dot{\mathbf{z}}\, \mathrm{d}v = \iiint\limits_{v_{\mathscr{P}}} \rho \mathbf{z} \times (\boldsymbol{\omega} \times \mathbf{z}) dv = \mathbf{J} \cdot \boldsymbol{\omega}, \tag{B.11}$$

where $\mathbf{v}_0 = \dot{\mathbf{u}}$ and $\mathbf{J}$ is the *inertia tensor*:

$$\mathbf{J} \triangleq \iiint\limits_{v_{\mathscr{P}}} \rho[(\mathbf{z} \cdot \mathbf{z})\mathbf{I} - \mathbf{z} \otimes \mathbf{z}]\, dv. \tag{B.12}$$

It is seen that $\mathbf{J}$ possesses the following property

$$\mathbf{J} = \mathbf{Q} \cdot \mathbf{J}_0 \cdot \mathbf{Q}^T, \qquad \mathbf{J}_0 \triangleq \iiint\limits_{V_{\mathscr{P}}} \rho[(\mathbf{Z} \cdot \mathbf{Z})\mathbf{I} - \mathbf{Z} \otimes \mathbf{Z}]\, dv, \tag{B.13}$$

where the volume integral is taken over $V_{\mathscr{P}}$ in the initial body configuration. The constant tensor $\mathbf{J}_0$ can be called the inertia tensor in the initial configuration. For example, for a solid homogeneous sphere of radius a, $\mathbf{J}$ is a spherical tensor

$$\mathbf{J} = \frac{2}{5} m a^2 \mathbf{I} = \mathbf{J}_0.$$

If the directors $\mathbf{d}_k$ are unit vectors along the principle axes of the inertia tensor, we see that $\mathbf{J}$ and $\mathbf{J}_0$ are diagonal

$$\mathbf{J} = J_1\mathbf{d}_1 \otimes \mathbf{d}_1 + J_2\mathbf{d}_2 \otimes \mathbf{d}_2 + J_3\mathbf{d}_3 \otimes \mathbf{d}_3,$$
$$\mathbf{J}_0 = J_1\mathbf{D}_1 \otimes \mathbf{D}_1 + J_2\mathbf{D}_2 \otimes \mathbf{D}_2 + J_3\mathbf{D}_3 \otimes \mathbf{D}_3,$$

where J_1, J_2, J_3 are moments of inertia with respect to the principal axes.

With regard to (B.8) and (B.13) it can be shown that the derivative of $\mathbf{J}$ satisfies the relation

$$\dot{\mathbf{J}} = \boldsymbol{\omega} \times \mathbf{J} - \mathbf{J} \times \boldsymbol{\omega}. \tag{B.14}$$

Taking the mass center as a pole we can rewrite the kinetic energy of the rigid body as follows

$$K \triangleq \frac{1}{2}\iiint\limits_{v_{\mathscr{P}}} \rho\mathbf{v}\cdot\mathbf{v}\,\mathrm{d}v = \frac{1}{2}m\mathbf{v}_0\cdot\mathbf{v}_0 + \frac{1}{2}\boldsymbol{\omega}\cdot\mathbf{J}\cdot\boldsymbol{\omega}. \tag{B.15}$$

The following identities are valid

$$\mathfrak{P} = \frac{\partial K}{\partial \mathbf{v}_0}, \qquad \mathfrak{M} = \frac{\partial K}{\partial \boldsymbol{\omega}}. \tag{B.16}$$

To a rigid body we can apply the forces and torques (couples or moments). The forces relates with the translation of the body whereas the torques involve body rotation.

The rigid body motion is described by two Euler's laws of motion.

1. **Balance of momentum. First Euler's law of motion of the rigid body.** *The time rate of the rigid body momentum is equal to the resultant vector of forces $\mathfrak{F}$, acting on the body:*

$$\frac{\mathrm{d}}{\mathrm{d}t}\mathfrak{P} = \mathfrak{F}, \qquad \mathfrak{F} \triangleq \iiint\limits_{v_{\mathscr{P}}} \rho\mathbf{f}\,\mathrm{d}v. \tag{B.17}$$

2. **Balance of moment of momentum. Second Euler's law of motion of the rigid body.** *The time rate of the rigid body moment of momentum with respect to pole o is equal to the resultant moment of all forces with respect to the pole and the body moments:*

$$\frac{\mathrm{d}}{\mathrm{d}t}\mathfrak{M} = \mathfrak{C}, \qquad \mathfrak{C} \triangleq \iiint\limits_{v_{\mathscr{P}}} \rho[(\mathbf{r} - \mathbf{r}_0) \times \mathbf{f} + \mathbf{m}]\mathrm{d}v. \tag{B.18}$$

Here $\mathbf{f}$ and $\mathbf{m}$ are the densities of the forces and the moments acting on the body, respectively.

In equilibrium, these laws reduce to the equality to zero of the resultant vector of the forces and the resultant moment:

$$\mathfrak{F} = \mathbf{0}, \qquad \mathfrak{C} = \mathbf{0}. \tag{B.19}$$

Substituting (B.11) to (B.18) and taking account (B.14), we get the motion equations of the rigid body

$$m\dot{\mathbf{v}}_0 = \mathfrak{F}, \quad \mathbf{J}\cdot\dot{\boldsymbol{\omega}} + \boldsymbol{\omega} \times \mathbf{J}\cdot\boldsymbol{\omega} = \mathfrak{C}. \tag{B.20}$$

Naturally, Eqs. (B.20) contain the n mass point motion equations (B.1) and (B.2) as a particular case.

Taking the principal axes of the inertia tensor $\mathbf{J}$ as the basis, we transform Eqs. (B.20) to the form

$$\begin{aligned} m\dot{v}_1 = \mathfrak{F}_1, \quad m\dot{v}_2 &= \mathfrak{F}_2, \quad m\dot{v}_3 = \mathfrak{F}_3, \\ J_1\dot{\omega}_1 + (J_3 - J_2)\omega_2\omega_3 &= \mathfrak{C}_1, \\ J_2\dot{\omega}_2 + (J_1 - J_3)\omega_1\omega_3 &= \mathfrak{C}_2, \\ J_3\dot{\omega}_3 + (J_2 - J_1)\omega_1\omega_2 &= \mathfrak{C}_3. \end{aligned}$$

(B.20) are simultaneous nonlinear ordinary differential equations. Only particular cases of analytic solutions are known for some relations between J_i, the solutions are due to Euler, Lagrange and Kovalevskaya[1]. In classical mechanics, the rigid body motion is also considered under some constraints like in the theory of rotation of a rigid body about an immovable point or axis, cf. for example [11]. It is worth of noting that rigid body motions are quite complicated. Many areas of modern mathematics arose from rigid body dynamics, cf. [12].

When displacements $\mathbf{u}$ and rotations $\boldsymbol{\theta}$ are infinitesimal we get simplified relations:

$$\mathbf{Q} \approx \mathbf{I} - \boldsymbol{\theta} \times \mathbf{I}, \quad \boldsymbol{\omega} = \dot{\boldsymbol{\theta}}.$$

Now Eqs. (B.20) reduce to a linear system of ordinary differential equations with respect to $\mathbf{u}$ and $\boldsymbol{\theta}$:

$$m\ddot{\mathbf{u}} = \mathfrak{F}, \quad \mathbf{J}_0\cdot\ddot{\boldsymbol{\theta}} = \mathfrak{C}. \tag{B.21}$$

In mechanics, Euler's laws are known since ancient time. For example, Eq. (B.19) was formulated as the lever law by Archimedes in the 3rd century BC. Euler showed that the second dynamic law, or the second equilibrium condition, is independent of the first one. From (B.18) and (B.19) we can derive Newton's laws for the system of mass points. As was mentioned in [13], to derive Euler's equations from Newton's law, we should introduce additional assumptions on the nature of the interaction between the points. Being more general, Eqs. (B.18) and (B.19) constitute the foundation of classic mechanics as well as of continuum mechanics.

---

[1] In the original paper "Sur le probleme de la rotation d'un corps solide autour d'un point fixe", Acta Mathematica, 1889, 12(1), 177–232, instead of Sofia Kovalevskaya the author was indicated as Sophie Kowalevski.

Appendix C
Elements of Mechanics of Elastic Rods

Mechanics of rods, beams and similar bodies one of dimensions of which is significantly bigger than two others has a long-time development. Nowadays, rod and beam structures are widely used in engineering, see [14, 15]. There are various approaches to derive governing equations for rods. First is the so-called direct approach used by Leonard Euler (1744) in his famous study of the elastica problem. In this work, the rod is represented by an elastic curve whose constitutive equations describe the curve as an one-dimensional (1D) continuum. The model was generalized by Cosserat brothers [16], who considered the translations and rotations as independent variables. Now this model is known as the Cosserat deformable curve. Since Euler's time, the direct approach is applied to derive the basic equations of rod mechanics in many works, see e.g. the seminal papers by Ericksen and Truesdell [17], Green and Naghdi [18], Kafadar [19], Green et al. [20] and the books [21, 22]. Other approaches of derivation of one-dimensional equations of rods are based on the use of the thinness hypothesis applied to three-dimensional thin rod-like bodies. The derivations itself can be performed by the application of kinematical or stress hypotheses, the integration over the rod cross-section, or within mathematical techniques like variational and asymptotic analysis, see e.g. [14, 23–27].

We consider the rod as a deformable Cosserat curve whose, kinematics is described by the translation and rotation fields. We will employ the kinematical model of *directed curves*, presented in [21, 23, 28]. In this approach, the thin rod-like bodies are modeled as deformable curves endowed with a triad of vectors (also called directors) attached to every point. This triad of directors rotate rigidly during deformation and gives thus information about rotations of the rod cross-sections. In other words, we consider the rod as a 1D Cosserat or micropolar continuum.

Let us briefly present the kinematical model of directed curves. We denote by the curve $\mathscr{C}$ in the reference configuration $\varkappa$ and by s the material coordinate along $\mathscr{C}$.

V. A. Eremeyev et al., *Foundations of Micropolar Mechanics*,
SpringerBriefs in Continuum Mechanics,
DOI: 10.1007/978-3-642-28353-6,

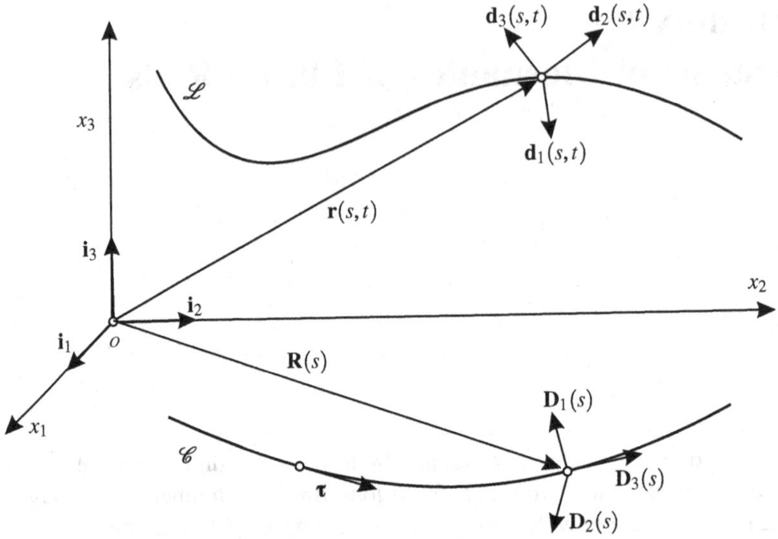

Fig. C.1 The actual and reference configurations of the rod

The coordinate s is taken to be the arclength parameter along $\mathscr{C}$. The reference configuration of the rod is defined by the vector fields (see Fig. C.1)

$$\mathbf{R} = \mathbf{R}(s), \qquad \mathbf{D}_k = \mathbf{D}_k(s), \quad k = 1,2,3, \quad s \in [0, l].$$

Here $\mathbf{R}$ is the position vector of points on $\mathscr{C}$, the triad $\{\mathbf{D}_1, \mathbf{D}_2, \mathbf{D}_3\}$ specifies the three unit vectors (directors) orthogonal to each other and l is the length of the rod. The vector $\mathbf{D}_3$ can be chosen tangent to the curve $\mathscr{C}$ (i.e. $\mathbf{D}_3 = \mathbf{R}' \equiv \boldsymbol{\tau}$), while $\mathbf{D}_1$ and $\mathbf{D}_2$ are usually taken along the principal axes of inertia of the rod cross-section. The prime designates the derivative with respect to the spatial coordinate s (i.e. $f' = \frac{\partial f}{\partial s}$).

Let χ denote the actual configuration at time t, which is determined by the position vector $\mathbf{r} = \mathbf{r}(s,t)$ of the curve $\mathscr{L}$ and the directors $\mathbf{d}_i = \mathbf{d}_i(s,t)$, $i = 1,2,3$. The three unit vectors $\mathbf{d}_i$ remain mutually orthogonal after deformation, but $\mathbf{d}_3$ is no longer tangent to the curve $\mathscr{L}$, which means that the model takes into account the transverse shear deformations of rods.

Differentiating $\mathbf{D}_k$ and $\mathbf{d}_k$ with respect to s we obtain the formulae

$$\mathbf{D}_k' = \mathbf{k}_0 \times \mathbf{D}_k, \quad \mathbf{d}_k' = \mathbf{k} \times \mathbf{d}_k. \tag{C.1}$$

Equations (C.1) are analogues of the Frenet–Serret formulae (A.38). The vectors $\mathbf{k}_0$ and $\mathbf{k}$ are analogues of the Darboux vector and can be called the *generalized Darboux vectors* in the reference and actual configurations, respectively.

We define the rotation tensor $\mathbf{Q}$ by the relation

$$\mathbf{Q}(s,t) = \mathbf{d}_k(s,t) \otimes \mathbf{D}_k(s). \tag{C.2}$$

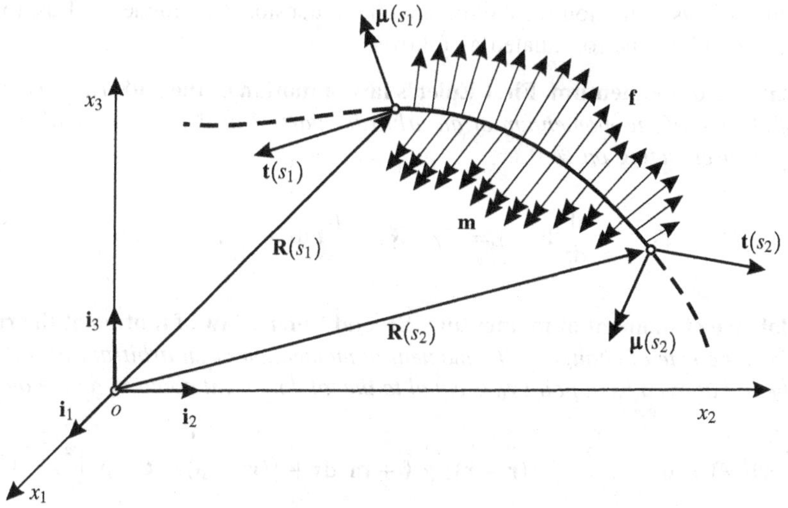

Fig. C.2 The part $\mathscr{P}$ of the rod and acting on $\mathscr{P}$ forces and couples

The velocity vector $\mathbf{v}$ is given by the formula $\mathbf{v}(s,t) = \dot{\mathbf{r}}(s,t)$ and the angular velocity vector $\boldsymbol{\omega}(s,t)$ is given by Eq. (B.9).

We introduce the kinetic energy density $K(s,t)$ by relation

$$K = \frac{1}{2}\rho(\mathbf{v}\cdot\mathbf{v} + 2\mathbf{v}\cdot\boldsymbol{\Theta}_1\cdot\boldsymbol{\omega} + \boldsymbol{\omega}\cdot\boldsymbol{\Theta}_2\cdot\boldsymbol{\omega}), \tag{C.3}$$

where ρ is the mass density in the reference configuration $\varkappa$, the second-order tensors $\boldsymbol{\Theta}_1(s,t)$ and $\boldsymbol{\Theta}_2(s,t)$ are the *tensors of inertia* (per unit mass), which characterize the distribution of the material in the cross-section. For the reference configuration, the tensors of inertia are denoted by $\boldsymbol{\Theta}_1^0(s)$ and $\boldsymbol{\Theta}_2^0(s)$. As in the case of rigid body dynamics, see (B.13), we have $\boldsymbol{\Theta}_\alpha = \mathbf{Q}\cdot\boldsymbol{\Theta}_\alpha^0\cdot\mathbf{Q}^T, \alpha = 1, 2$.

We assume the following formulae for the momentum $\mathfrak{P}$ and the moment of momentum $\mathfrak{M}$ of any part $\mathscr{P}$ of the rod

$$\mathfrak{P}(\mathscr{P}) \triangleq \int_{s_1}^{s_2} \rho\mathbf{K}_1 \mathrm{d}s, \quad \mathfrak{M}(\mathscr{P}) \triangleq \int_{s_1}^{s_2} \rho\{(\mathbf{r} - \mathbf{r}_0) \times \mathbf{K}_1 + \mathbf{K}_2\}\mathrm{d}s, \tag{C.4}$$

where $\mathbf{K}_1$ and $\mathbf{K}_2$ are given by

$$\mathbf{K}_1 \triangleq \frac{\partial K}{\partial \mathbf{v}} = \mathbf{v} + \boldsymbol{\Theta}_1^T\cdot\boldsymbol{\omega}, \quad \mathbf{K}_2 \triangleq \frac{\partial K}{\partial \boldsymbol{\omega}} = \boldsymbol{\Theta}_1\cdot\mathbf{v} + \boldsymbol{\Theta}_2\cdot\boldsymbol{\omega}.$$

$s_1, s_2 \in [0, l]$ are the coordinates of $\mathscr{P}$ in the reference configuration, see Fig. C.2.

Euler's laws of motion for the rod are one-dimensional analogues of Eqs. (3.3) and (3.4) and can be formulated as follows:

1. **Balance of momentum. First Euler's law of motion of the rod**. *The time rate of change of the momentum of an arbitrary part $\mathscr{P}$ of the rod is equal to the total force acting on $\mathscr{P}$:*

$$\frac{\mathrm{d}}{\mathrm{d}t}\mathfrak{P}(\mathscr{P}) = \mathfrak{F}, \quad \mathfrak{F} \triangleq \int\limits_{s_1}^{s_2} \mathbf{f}\,\mathrm{d}s + \mathbf{t}\Big|_{s_1}^{s_2}. \tag{C.5}$$

2. **Balance of moment of momentum. Second Euler's law of motion of the rod.** *The time rate of change of the moment of momentum of an arbitrary part $\mathscr{P}$ of the rod about a fixed point $\mathbf{r}_0$ is equal to the total moment about $\mathbf{r}_0$ acting on $\mathscr{P}$:*

$$\frac{\mathrm{d}}{\mathrm{d}t}\mathfrak{M}(\mathscr{P}) = \mathfrak{C}, \quad \mathfrak{C} \triangleq \int\limits_{s_1}^{s_2} \{(\mathbf{r}-\mathbf{r}_0)\times\mathbf{f}+\mathbf{m}\}\mathrm{d}s + \{(\mathbf{r}-\mathbf{r}_0)\times\mathbf{t}+\boldsymbol{\mu}\}\Big|_{s_1}^{s_2}. \tag{C.6}$$

Here $\mathbf{f}$ and $\mathbf{m}$ are the external resultant body force and moment per unit length of $\mathscr{C}$, $\mathbf{t}$ and $\boldsymbol{\mu}$ are the force and moment vectors, respectively.

Under suitable smoothness assumptions, we obtain the following local equations of the principles (C.5) and (C.6) that are the Lagrangian equations of motion:

$$\begin{aligned} &\mathbf{t}'(s,t)+\mathbf{f} = \rho\frac{\mathrm{d}}{\mathrm{d}t}(\mathbf{v}+\boldsymbol{\Theta}_1\cdot\boldsymbol{\omega}),\\ &\boldsymbol{\mu}'(s,t)+\mathbf{r}'\times\mathbf{t}(s,t)+\mathbf{m} = \rho\left[\frac{\mathrm{d}}{\mathrm{d}t}(\boldsymbol{\Theta}_1^T\cdot\mathbf{v}+\boldsymbol{\Theta}_2\cdot\boldsymbol{\omega})+\mathbf{v}\times\boldsymbol{\Theta}_1\cdot\boldsymbol{\omega}\right]. \end{aligned} \tag{C.7}$$

For an elastic rod there exists the strain energy density W. According to the equipresence principle we take W in the form $W = W(\mathbf{r},\mathbf{r}',\mathbf{Q},\mathbf{Q}')$. Applying to W the material frame-indifference principle we derive that W depends on two vectorial relative Lagrangian strain measures

$$W = W(\mathbf{E},\mathbf{K}), \qquad \mathbf{E} = \mathbf{Q}^T\cdot\mathbf{r}'-\boldsymbol{\tau}, \quad \mathbf{K} = -\frac{1}{2}\left(\mathbf{Q}^T\cdot\mathbf{Q}'\right)_\times = \mathbf{Q}^T\cdot\mathbf{k}-\mathbf{k}_0, \tag{C.8}$$

where $\mathbf{E}$ is the vector of stretching–shear and $\mathbf{K}$ is the vector of bending–twisting. We notice that (essentially) the same strain measures have been considered in other approaches to the rod theory [21, 23–25].

Using the virtual work principle we can derive the constitutive equations in the following form, see [23],

$$\mathbf{t} = \frac{\partial W}{\partial \mathbf{E}}\cdot\mathbf{Q}^T, \qquad \boldsymbol{\mu} = \frac{\partial W}{\partial \mathbf{K}}\cdot\mathbf{Q}^T. \tag{C.9}$$

Mechanics of elastic rods presented above gives a good example of an one-dimensional Cosserat continuum. Assuming certain constraints one can derive Timoshenko or Euler–Bernoulli models of rods [21–23]. On the other hand the theory of rods give also the “generalized” one-dimensional models of continuum which possess more than six independent kinematical degrees of freedom. Examples of such “generalized” models are the theory of thin-walled rods or theories taking into account warping of the rod cross-section [14, 15, 23].

Appendix D
Micropolar Plates and Shells as Two-Dimensional Cosserat Continua

Since the paper by Ericksen and Truesdell [17] the generalized models of shells and plates of Cosserat type are extensively discussed in the literature, see the recent review [29].

In what follows we consider the micropolar shell model as an example of the 2D Cosserat continuum. Indeed, a micropolar shell is a two-dimensional analogue of the three-dimensional micropolar continuum, i.e. a micropolar shell is a deformable directed surface each particle of which has six degrees of freedom as a rigid body. The micropolar shell kinematics is described by two fields. The first field is given by the position vector of the base surface of the shell while the second one is the proper orthogonal tensor that describes the rotation of a shell cross-section. In contrast to Kirchhoff–Love and Mindlin–Reissner's theories of plates and shells, the boundary-value problems for a micropolar shell consist of six scalar equations supplemented with six boundary conditions. Within the micropolar shell theory, the so-called drilling moment can be taken into account.

In [30–33] the basic equations of the micropolar shells theory were derived with using the direct approach. They coincide with the equations of the general nonlinear shell theory initiated by Reissner [34] and presented by Libai and Simmonds [35, 36], Pietraszkiewicz [37], and Chróścielewski et al. [38]. The micropolar shell theory is also named the six-parameter theory of shells. Numerous examples of FEM-calculations in the frame of micropolar shell theory can be found in [38–41]. The model of thin-walled structures made of the materials undergoing stress-induced phase transitions is presented in [42–45]. The thermodynamical extensions of the mechanics of micropolar shells are discussed in [32, 43, 44, 46–48].

V. A. Eremeyev et al., *Foundations of Micropolar Mechanics*,
SpringerBriefs in Continuum Mechanics,
DOI: 10.1007/978-3-642-28353-6,

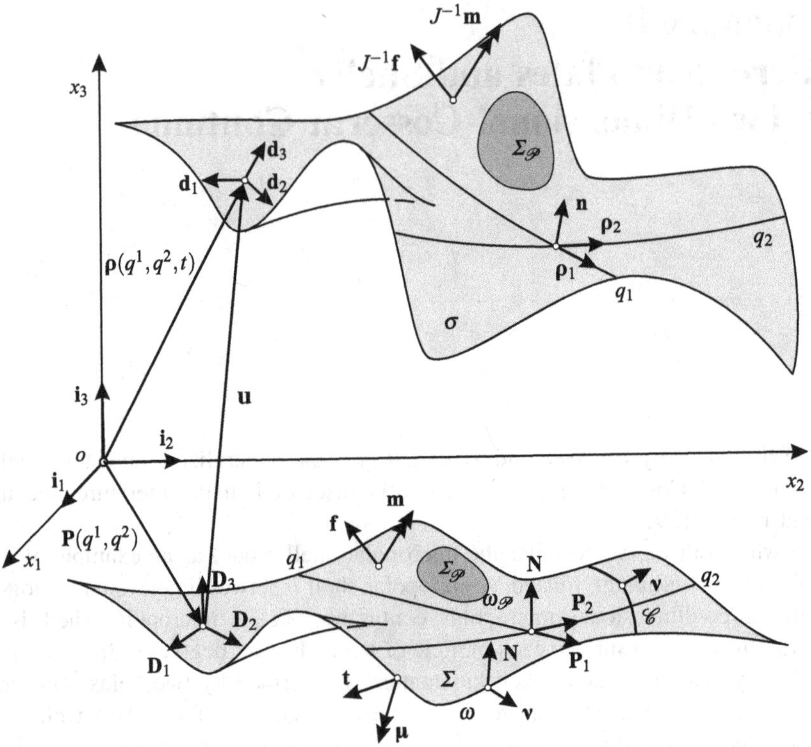

Fig. D.1 Kinematics of a micropolar shell

D.1 Kinematics of a Micropolar Shell

Let Σ be a base surface of the micropolar shell in the reference configuration $\varkappa$ (for example, in the undeformed shell state), $q^\alpha (\alpha = 1, 2)$ Gaussian coordinates on Σ, and $\mathbf{P}(q^1, q^2)$ the position vector of the points of Σ, see Fig. D.1. In the actual, deformed, configuration χ the base surface is denoted by σ, and the position of its material points (infinitesimal point-bodies) is given by vector $\boldsymbol{\rho}(q^1, q^2)$. The point-body orientation is described by the *microrotation tensor* $\mathbf{Q}(q^1, q^2)$ that is a proper orthogonal tensor. Introducing three orthonormal vectors $\mathbf{D}_k \, (k = 1, 2, 3)$ describing the orientation in the reference configuration, and three orthonormal vectors $\mathbf{d}_k$ determining the orientation in the actual configuration, we get tensor $\mathbf{Q}$ in the form $\mathbf{Q} = \mathbf{d}_k \otimes \mathbf{D}_k$. Thus the micropolar shell is described by two kinematically independent fields

$$\boldsymbol{\rho} = \boldsymbol{\rho}(q^\alpha) \quad \text{and} \quad \mathbf{Q} = \mathbf{Q}(q^\alpha). \tag{D.1}$$

For a micropolar hyper-elastic shell we can introduce a strain energy density W. With regard for the local action principle [8], W takes the form

$$W = W(\boldsymbol{\rho}, \nabla_s\boldsymbol{\rho}, \mathbf{Q}, \nabla_s\mathbf{Q}),$$

where

$$\nabla_s\boldsymbol{\psi} \triangleq \mathbf{P}^\alpha \otimes \frac{\partial\boldsymbol{\psi}}{\partial q^\alpha} \quad (\alpha, \beta = 1, 2), \quad \mathbf{P}^\alpha \cdot \mathbf{P}_\beta = \delta^\alpha_\beta, \quad \mathbf{P}^\alpha \cdot \mathbf{N} = 0, \quad \mathbf{P}_\beta = \frac{\partial \mathbf{P}}{\partial q^\beta}.$$

Here vectors $\mathbf{P}_\beta$ and $\mathbf{P}^\alpha$ denote the natural and reciprocal bases on Σ respectively, $\mathbf{N}$ is the unit normal to Σ, ∇_s is the surface nabla operator on Σ, and $\boldsymbol{\psi}$ is an arbitrary differentiable tensor field given on Σ.

From the principle of material frame-indifference we can deduce that W depends on two surface strain measures $\mathbf{E}$ and $\mathbf{K}$ of Cosserat type:

$$W = W(\mathbf{E}, \mathbf{K}),$$

where

$$\mathbf{E} = \mathbf{F}\cdot\mathbf{Q}^T - \mathbf{A}, \quad \mathbf{K} = \frac{1}{2}\mathbf{P}^\alpha \otimes \left(\frac{\partial \mathbf{Q}}{\partial q^\alpha}\cdot\mathbf{Q}^T\right)_\times, \quad \mathbf{F} = \nabla_s\boldsymbol{\rho}. \tag{D.2}$$

Here $\mathbf{F}$ is the surface deformation gradient, and $\mathbf{A} \triangleq \mathbf{I} - \mathbf{N} \otimes \mathbf{N}$.

The proper orthogonal tensor describing the rotation about axis $\mathbf{e}$ for angle φ can be represented with use of Gibbs's formula (A.7). Introducing the finite rotation vector $\boldsymbol{\theta} = 2\mathbf{e}\tan\varphi/2$ we get a representation of $\mathbf{Q}$ in the form (A.8) that does not contain trigonometric functions. By Eq. (A.8), a proper orthogonal tensor $\mathbf{Q}$ defines uniquely vector $\boldsymbol{\theta}$

$$\boldsymbol{\theta} = 2(1 + \operatorname{tr}\mathbf{Q})^{-1}\mathbf{Q}_\times. \tag{D.3}$$

Using the *finite rotation vector* $\boldsymbol{\theta}$ we can express $\mathbf{K}$ as follows

$$\mathbf{K} = \mathbf{P}^\alpha \otimes \mathbf{L}_\alpha = \frac{4}{4 + \theta^2}\nabla_s\boldsymbol{\theta}\cdot\left(\mathbf{i} + \frac{1}{2}\mathbf{I} \times \boldsymbol{\theta}\right). \tag{D.4}$$

Equation (D.4) is a particular case of (4.33). The strain measures $\mathbf{E}$ and $\mathbf{K}$ are two-dimensional analogues of the strain measures used in 3D Cosserat continuum, see Sect. 4.2.

D.2 The Virtual Work Principle and Formulation of Boundary Value Problems

Lagrangian equilibrium equations for a micropolar shell can be derived from *the virtual work principle*

$$\delta \iint_\Sigma W \,\mathrm{d}\Sigma = \delta' A, \tag{D.5}$$

where

$$\delta' A = \iint\limits_{\Sigma} (\mathbf{f}\cdot\delta\boldsymbol{\rho} + \mathbf{m}\cdot\delta'\boldsymbol{\psi})\mathrm{d}\Sigma + \int\limits_{\omega_2} \mathbf{t}\cdot\delta\boldsymbol{\rho}\,\mathrm{d}s + \int\limits_{\omega_4} \boldsymbol{\mu}\cdot\delta'\boldsymbol{\psi}\,\mathrm{d}s, \quad \mathbf{I}\times\delta'\boldsymbol{\psi} = -\mathbf{Q}^T\cdot\delta\mathbf{Q}.$$

In Eq. (D.5), δ is the symbol of variation, $\delta'\boldsymbol{\psi}$ the virtual rotation vector, $\mathbf{f}$ the surface force density distributed on Σ, and $\mathbf{m}$ the surface couple density distributed on Σ. The quantities $\mathbf{t}$ and $\boldsymbol{\mu}$ are the linear densities of forces and couples distributed along corresponding parts of the shell boundary ω, respectively.

Using the formulae from [31,32],

$$\delta W = \frac{\partial W}{\partial \mathbf{E}} \bullet \delta\mathbf{E} + \frac{\partial W}{\partial \mathbf{K}} \bullet \delta\mathbf{K},$$
$$\delta\mathbf{E} = (\nabla_s\delta\boldsymbol{\rho})\cdot\mathbf{Q}^T + \mathbf{F}\cdot\delta\mathbf{Q}^T, \delta\mathbf{K} = (\nabla_s\delta'\boldsymbol{\psi})\cdot\mathbf{Q}^T, \delta'\boldsymbol{\psi} = \frac{4}{4+\theta^2}\left(\delta\boldsymbol{\theta} + \frac{1}{2}\boldsymbol{\theta}\times\delta\boldsymbol{\theta}\right),$$

and Eq. (D.5), we obtain the *Lagrangian shell equations*:

$$\nabla_s\cdot\mathbf{D} + \mathbf{f} = \mathbf{0}, \quad \nabla_s\cdot\mathbf{G} + \left[\mathbf{F}^T\cdot\mathbf{D}\right]_\times + \mathbf{m} = \mathbf{0}, \tag{D.6}$$

$$\mathbf{D} = \mathbf{P}_1\cdot\mathbf{Q}, \quad \mathbf{G} = \mathbf{P}_2\cdot\mathbf{Q}, \qquad \mathbf{P}_1 = \frac{\partial W}{\partial \mathbf{E}}, \quad \mathbf{P}_2 = \frac{\partial W}{\partial \mathbf{K}}, \tag{D.7}$$

They are supplemented by the boundary conditions:

$$\begin{aligned} &\text{on } \omega_1 : \boldsymbol{\rho} = \boldsymbol{\rho}(s), \quad \text{on } \omega_2 : \boldsymbol{\nu}\cdot\mathbf{D} = \mathbf{t}(s), \\ &\text{on } \omega_3 : \mathbf{Q} = \mathbf{h}(s), \mathbf{h}\cdot\mathbf{h}^T = \mathbf{I}, \quad \text{on } \omega_4 : \boldsymbol{\nu}\cdot\mathbf{G} = \boldsymbol{\mu}(s). \end{aligned} \tag{D.8}$$

Here $\boldsymbol{\rho}(s)$, $\mathbf{h}(s)$ are given vector and tensor functions, and $\boldsymbol{\nu}$ is the external unit normal to the boundary curve ω ($\boldsymbol{\nu}\cdot\mathbf{N} = 0$). Equations (D.6) are the equilibrium equations for the linear momentum and angular momentum at any shell point. $\mathbf{D}$ and $\mathbf{G}$ are *the surface stress and couple stress tensors* of the first Piola–Kirchhoff type and the stress measures $\mathbf{P}_1$ and $\mathbf{P}_2$ in Eqs. (D.6) are the referential stress and couple stress tensors, respectively. The strain measures $\mathbf{E}$ and $\mathbf{K}$ are work-conjugate to the stress measures $\mathbf{D}$ and $\mathbf{G}$. The boundary ω of Σ is divided into two parts in such a way that $\omega = \omega_1 \cup \omega_2 = \omega_3 \cup \omega_4$.

The following relations are valid: $\mathbf{N}\cdot\mathbf{D} = \mathbf{N}\cdot\mathbf{G} = \mathbf{N}\cdot\mathbf{P}_1 = \mathbf{N}\cdot\mathbf{P}_2 = \mathbf{0}$.

The *equilibrium equations* (D.6) may be transformed to the *Eulerian form*

$$\tilde{\nabla}_s\cdot\mathbf{T} + J^{-1}\mathbf{f} = \mathbf{0}, \quad \tilde{\nabla}_s\cdot\mathbf{M} + \mathbf{T}_\times + J^{-1}\mathbf{m} = \mathbf{0}, \tag{D.9}$$

where

$$\tilde{\nabla}_s\cdot\boldsymbol{\psi} \triangleq \boldsymbol{\rho}^\alpha\cdot\frac{\partial\boldsymbol{\psi}}{\partial q^\alpha}, \quad \boldsymbol{\rho}^\alpha\cdot\boldsymbol{\rho}_\beta = \delta^\alpha_\beta, \boldsymbol{\rho}^\alpha\cdot\mathbf{n} = 0, \quad \boldsymbol{\rho}_\beta = \frac{\partial\boldsymbol{\rho}}{\partial q^\beta},$$
$$\mathbf{T} = J^{-1}\mathbf{F}^T\cdot\mathbf{D}, \quad \mathbf{M} = J^{-1}\mathbf{F}^T\cdot\mathbf{G}, \tag{D.10}$$

$$J = \sqrt{\frac{1}{2}\left\{\left[\mathrm{tr}\left(\mathbf{F}\cdot\mathbf{F}^T\right)\right]^2 - \mathrm{tr}\left[\left(\mathbf{F}\cdot\mathbf{F}^T\right)^2\right]\right\}}.$$

Here $\mathbf{T}$ and $\mathbf{M}$ are Cauchy-type surface stress and couple stress tensors, $\tilde{\nabla}_s$ is the surface nabla operator on σ related with ∇_s by the formula $\nabla_s = \mathbf{F}\cdot\tilde{\nabla}_s$, and $\mathbf{n}$ is the unit normal to σ.

Under some natural restrictions, the equilibrium problem for a micropolar shell can be transformed to the system with respect to the strain measures:

$$\nabla_s\cdot\mathbf{P}_1 - \left(\mathbf{P}_1^T\cdot\mathbf{K}\right)_\times + \mathbf{f}^* = \mathbf{0}; \quad \nabla_s\cdot\mathbf{P}_2 - \left(\mathbf{P}_2^T\cdot\mathbf{K} + \mathbf{P}_1^T\cdot\mathbf{E}\right)_\times + \mathbf{m}^* = \mathbf{0}, \qquad \text{(D.11)}$$

$$\omega_2 \ : \ \mathbf{v}\cdot\mathbf{P}_1 = \mathbf{t}^*, \qquad \omega_4 \ : \ \mathbf{v}\cdot\mathbf{P}_2 = \boldsymbol{\mu}^*, \qquad \text{(D.12)}$$

$$\mathbf{f}^* \triangleq \mathbf{f}\cdot\mathbf{Q}^T, \quad \mathbf{m}^* \triangleq \mathbf{m}\cdot\mathbf{Q}^T, \quad \mathbf{t}^* \triangleq \mathbf{t}\cdot\mathbf{Q}^T, \quad \boldsymbol{\mu}^* \triangleq \boldsymbol{\mu}\cdot\mathbf{Q}^T.$$

Let the vectors $\mathbf{f}^*$, $\mathbf{m}^*$, and $\mathbf{t}^*, \boldsymbol{\mu}^*$ be given as some functions of coordinates q^1, q^2, and s. From the physical point of view, it means that the shell is loaded by tracking forces and couples. Then Eqs. (D.11)–(D.12) depend on $\mathbf{E}$, $\mathbf{K}$ that are the only independent fields.

The *motion equations* of the micropolar shell are (cf. for example [31, 36, 38])

$$\nabla_s\cdot\mathbf{D} + \mathbf{f} = \rho\frac{\mathrm{d}\mathbf{K}_1}{\mathrm{d}t}, \quad \nabla_s\cdot\mathbf{G} + \left[\mathbf{F}^T\cdot\mathbf{D}\right]_\times + \mathbf{m} = \rho\left(\frac{\mathrm{d}\mathbf{K}_2}{\mathrm{d}t} + \mathbf{v}\times\boldsymbol{\Theta}_1^T\cdot\boldsymbol{\omega}\right), \qquad \text{(D.13)}$$

where

$$\mathbf{K}_1 \triangleq \frac{\partial K}{\partial \mathbf{v}} = \mathbf{v} + \boldsymbol{\Theta}_1^T\cdot\boldsymbol{\omega}, \quad \mathbf{K}_2 \triangleq \frac{\partial K}{\partial \boldsymbol{\omega}} = \boldsymbol{\Theta}_1\cdot\mathbf{v} + \boldsymbol{\Theta}_2\cdot\boldsymbol{\omega}, \qquad \text{(D.14)}$$

$$K(\mathbf{v}, \boldsymbol{\omega}) = \frac{1}{2}\mathbf{v}\cdot\mathbf{v} + \boldsymbol{\omega}\cdot\boldsymbol{\Theta}_1\cdot\mathbf{v} + \frac{1}{2}\boldsymbol{\omega}\cdot\boldsymbol{\Theta}_2\cdot\boldsymbol{\omega}. \qquad \text{(D.15)}$$

Here

$$\mathbf{v} = \frac{\mathrm{d}\boldsymbol{\rho}}{\mathrm{d}t}, \qquad \boldsymbol{\omega} = \frac{1}{2}\left(\mathbf{Q}^T\cdot\frac{\mathrm{d}\mathbf{Q}}{\mathrm{d}t}\right)_\times$$

are the linear and angular velocities, respectively, ρ is the surface mass density in the reference configuration, ρK is the surface density of the kinetic energy, and $\rho\boldsymbol{\Theta}_1$, $\rho\boldsymbol{\Theta}_2$ are *the rotatory inertia tensors* ($\boldsymbol{\Theta}_2^T = \boldsymbol{\Theta}_2$). For the dynamic problem (D.13), the initial conditions are

$$\boldsymbol{\rho}\big|_{t=0} = \boldsymbol{\rho}^\circ, \quad \mathbf{v}\big|_{t=0} = \mathbf{v}^\circ, \quad \mathbf{Q}\big|_{t=0} = \mathbf{Q}^\circ, \quad \boldsymbol{\omega}\big|_{t=0} = \boldsymbol{\omega}^\circ,$$

with given initial values $\boldsymbol{\rho}^\circ$, $\mathbf{v}^\circ$, $\mathbf{Q}^\circ$, $\boldsymbol{\omega}^\circ$.

D.3 Euler's Motion Laws of a Micropolar Shell

The surface stress measures and the motion equations can be introduced using two-dimensional analogues of Euler's 3D motion laws. Let us define the momentum $\mathfrak{P}$ and moment of momentum $\mathfrak{M}$ of a shell part $\mathscr{P}$ as follows

$$\mathfrak{P}(\mathscr{P}) \triangleq \iint\limits_{\Sigma_{\mathscr{P}}} \rho \mathbf{K}_1 \mathrm{d}\Sigma, \quad \mathfrak{M}(\mathscr{P}) \triangleq \iint\limits_{\Sigma_{\mathscr{P}}} \rho\{(\boldsymbol{\rho} - \boldsymbol{\rho}_0) \times \mathbf{K}_1 + \mathbf{K}_2\}\mathrm{d}\Sigma, \tag{D.16}$$

where $\mathbf{K}_1$ and $\mathbf{K}_2$ are determined by (D.14), $\Sigma_{\mathscr{P}} \subset \Sigma$ is the part of Σ corresponding to $\mathscr{P}$ in the reference configuration, see Fig. D.1. Equations (D.14) are the kinetic constitutive equations of the micropolar shell. A more general form of the kinetic constitutive equations is discussed by Pietraszkiewicz [48].

Euler's motion laws for the shell are analogues of Eqs. (3.3) and (3.4), they are formulated as follows:

1. **Balance of momentum. First Euler's law of motion of the shell.** *The time rate of change of the momentum of an arbitrary shell part $\mathscr{P}$ is equal to the total force acting on $\mathscr{P}$:*

$$\frac{\mathrm{d}}{\mathrm{d}t}\mathfrak{P}(\mathscr{P}) = \mathfrak{F}, \quad \mathfrak{F} \triangleq \iint\limits_{\Sigma_{\mathscr{P}}} \mathbf{f}\,\mathrm{d}\Sigma + \int\limits_{\omega_{\mathscr{P}}} \mathbf{t}\,\mathrm{d}\omega. \tag{D.17}$$

2. **Balance of moment of momentum. Second Euler's law of motion of the shell.** *The time rate of change of the moment of momentum of an arbitrary shell part $\mathscr{P}$ about a fixed point $\boldsymbol{\rho}_0$ is equal to the total moment about $\boldsymbol{\rho}_0$ acting on $\mathscr{P}$:*

$$\frac{\mathrm{d}}{\mathrm{d}t}\mathfrak{M}(\mathscr{P}) = \mathfrak{C}, \quad \mathfrak{C} \triangleq \iint\limits_{\Sigma_{\mathscr{P}}} \{(\boldsymbol{\rho} - \boldsymbol{\rho}_0) \times \mathbf{f} + \mathbf{m}\}\mathrm{d}\Sigma + \int\limits_{\omega_{\mathscr{P}}} \{(\boldsymbol{\rho} - \boldsymbol{\rho}_0) \times \mathbf{t} + \boldsymbol{\mu}\}\mathrm{d}\omega. \tag{D.18}$$

As for 3D Cosserat continuum, using (D.17) and (D.18) we can prove two-dimensional analogues to the Cauchy lemma and Cauchy theorem and afterwards introduce the surface stress measures and derive the motion equations of a micropolar shell.

D.4 Constitutive Equations for a Micropolar Shell

For an elastic shell, the constitutive equations are defined by the surface strain energy density as the function of two strain measures. An example is presented by the model of a *physically linear isotropic shell*, see [31, 38, 49], whose energy is given by the quadratic form

$$
\begin{aligned}
2W &= \alpha_1 \mathrm{tr}^2 \mathbf{E}_{\parallel} + \alpha_2 \mathrm{tr}\, \mathbf{E}_{\parallel}^2 + \alpha_3 \mathrm{tr}\left(\mathbf{E}_{\parallel} \cdot \mathbf{E}_{\parallel}^T\right) + \alpha_4 \mathbf{N} \cdot \mathbf{E}^T \cdot \mathbf{E} \cdot \mathbf{N} \\
&+ \beta_1 \mathrm{tr}^2 \mathbf{K}_{\parallel} + \beta_2 \mathrm{tr}\, \mathbf{K}_{\parallel}^2 + \beta_3 \mathrm{tr}\left(\mathbf{K}_{\parallel} \cdot \mathbf{K}_{\parallel}^T\right) + \beta_4 \mathbf{N} \cdot \mathbf{K}^T \cdot \mathbf{K} \cdot \mathbf{N},
\end{aligned}
\tag{D.19}
$$

where $\mathbf{E}_{\parallel} \triangleq \mathbf{E} \cdot \mathbf{A}$, $\mathbf{K}_{\parallel} \triangleq \mathbf{K} \cdot \mathbf{A}$. In Eq. (D.19) there is no term that is bilinear in $\mathbf{E}$ and $\mathbf{K}$, it is a consequence of the fact that the surface wryness tensor $\mathbf{K}$ is a axial tensor that changes the sign on a space mirror reflection. Note the constitutive equations contain 8 parameters, α_k, β_k $k = 1, 2, 3, 4$.

For the elastic moduli in Eq. (D.19) in [38] there were used the relations:

$$
\alpha_1 = C\nu, \quad \alpha_2 = 0, \quad \alpha_3 = C(1-\nu), \quad \alpha_4 = \alpha_s C(1-\nu), \quad C = \frac{Eh}{1-\nu^2},
$$

$$
\beta_1 = D\nu, \quad \beta_2 = 0, \quad \beta_3 = D(1-\nu), \quad \beta_4 = \alpha_t D(1-\nu), \quad D = \frac{Eh^3}{12(1-\nu^2)},
$$

where E is the Young modulus, ν is the Poisson ratio of the bulk material, α_s and α_t are dimensionless shear correction factors, and h is the shell thickness. Parameter α_s is the shear correction factor. Reissner [50] used $\alpha_s = 5/6$ in his plate theory, by Mindlin [51] $\alpha_s = \pi^2/12$. For the couple stresses parameter α_t plays a role similar to α_s for the stresses. The value $\alpha_t = 0.7$ was proposed by Pietraszkiewicz [52, 53], also see [39]. In [38–40] the influence of α_s and α_t on the solution is investigated numerically for few boundary value problems.

For some types of anisotropy, other representations of shell energy density W were constructed [49] using material symmetry groups.

D.5 Linear Theory of Micropolar Shells

For small strains the shell equations can be simplified significantly. In geometrically linear version, Eulerian and Lagrangian shell descriptions do not differ as the difference between σ and Σ is considered to be infinitesimal. Here we do not distinguish the operators $\tilde{\nabla}_s$ and ∇_s as well as the types of stress and couple stress tensors in different configurations.

Let us introduce the *vector of infinitesimal translations* $\mathbf{u}$ and the *vector of infinitesimal rotations* $\boldsymbol{\vartheta}$ such that

$$
\boldsymbol{\rho} \approx \mathbf{R} + \mathbf{u}, \quad \mathbf{Q} \approx \mathbf{I} - \mathbf{I} \times \boldsymbol{\vartheta}. \tag{D.20}
$$

The formula for $\mathbf{Q}$ follows from the representation of a proper orthogonal tensor through the finite rotation vector (A.8) for $|\boldsymbol{\theta}| \ll 1$.

The stretch measure $\mathbf{E}$ and the wryness tensor $\mathbf{K}$ can be expressed in terms of the *linear stretch tensor* $\boldsymbol{\varepsilon}$ and the *linear wryness tensor* $\boldsymbol{\kappa}$ up to a linear addendum:

$$
\mathbf{E} \approx \mathbf{I} + \boldsymbol{\varepsilon}, \quad \mathbf{K} \approx \boldsymbol{\kappa}, \quad \boldsymbol{\varepsilon} = \nabla_s \mathbf{u} + \mathbf{A} \times \boldsymbol{\vartheta}, \quad \boldsymbol{\kappa} = \nabla_s \boldsymbol{\vartheta}. \tag{D.21}
$$

$\boldsymbol{\varepsilon}$ and $\boldsymbol{\varkappa}$ are used in the linear theory of micropolar shells, cf. [1, 31–33, 38]. As a consequence of (D.21) in the linear shell theory, the stress tensors $\mathbf{D}$, $\mathbf{P}_1$, and $\mathbf{T}$ coincide, the couple tensors $\mathbf{G}$, $\mathbf{P}_2$, $\mathbf{M}$ do not differ as well. In what follows, we will denote the stress tensor by $\mathbf{T}$ and the couple stress tensor by $\mathbf{M}$.

For a linearly elastic shell, the constitutive equations can be represented through the strain energy density $W = W(\boldsymbol{\varepsilon}, \boldsymbol{\kappa})$ as it follows

$$\mathbf{T} = \frac{\partial W}{\partial \boldsymbol{\varepsilon}}, \quad \mathbf{M} = \frac{\partial W}{\partial \boldsymbol{\kappa}}. \tag{D.22}$$

The equilibrium equations in the linear theory are

$$\nabla_s \cdot \mathbf{T} + \mathbf{f} = \mathbf{0}, \quad \nabla_s \cdot \mathbf{M} + \mathbf{T}_\times + \mathbf{m} = \mathbf{0}, \tag{D.23}$$

whereas the boundary conditions transform to

$$\begin{aligned} \text{on } \omega_1: \quad \mathbf{u} &= \mathbf{u}_0(s), \quad \text{on } \omega_2: \quad \boldsymbol{\nu}\cdot\mathbf{T} = \mathbf{t}(s), \\ \text{on } \omega_3: \quad \boldsymbol{\vartheta} &= \boldsymbol{\vartheta}_0(s), \quad \text{on } \omega_4: \quad \boldsymbol{\nu}\cdot\mathbf{M} = \boldsymbol{\mu}(s), \end{aligned} \tag{D.24}$$

where $\mathbf{u}_0(s)$ and $\boldsymbol{\vartheta}_0(s)$ are given functions of the arclength s; the conditions define the translations and rotations on contour parts ω_k.

For small strains, an example of constitutive equations is defined by the following quadratic form

$$\begin{aligned} 2W = {} & \alpha_1 \mathrm{tr}^2 \boldsymbol{\varepsilon}_\parallel + \alpha_2 \mathrm{tr}\, \boldsymbol{\varepsilon}_\parallel^2 + \alpha_3 \mathrm{tr}\left(\boldsymbol{\varepsilon}_\parallel \cdot \boldsymbol{\varepsilon}_\parallel^T\right) + \alpha_4 \mathbf{N}\cdot\boldsymbol{\varepsilon}^T\cdot\boldsymbol{\varepsilon}\cdot\mathbf{N} \\ & + \beta_1 \mathrm{tr}^2 \boldsymbol{\kappa}_\parallel + \beta_2 \mathrm{tr}\, \boldsymbol{\kappa}_\parallel^2 + \beta_3 \mathrm{tr}\left(\boldsymbol{\kappa}_\parallel \cdot \boldsymbol{\kappa}_\parallel^T\right) + \beta_4 \mathbf{N}\cdot\boldsymbol{\kappa}^T\cdot\boldsymbol{\kappa}\cdot\mathbf{N} \end{aligned} \tag{D.25}$$

that describes a *linear isotropic shell*. Here α_k and β_k, $k = 1, 2, 3, 4$, are elastic constants, and $\boldsymbol{\varepsilon}_\parallel \triangleq \boldsymbol{\varepsilon}\cdot\mathbf{A}$, $\boldsymbol{\kappa}_\parallel \triangleq \boldsymbol{\kappa}\cdot\mathbf{A}$.

By Eqs. (D.22) and (D.25), the stress tensor and the couple stress tensor are

$$\mathbf{T} = \alpha_1 \mathbf{A} \mathrm{tr}\, \boldsymbol{\varepsilon}_\parallel + \alpha_2 \epsilon_\parallel^T + \alpha_3 \boldsymbol{\varepsilon}_\parallel + \alpha_4 \boldsymbol{\varepsilon}\cdot\mathbf{N}\otimes\mathbf{N}, \tag{D.26}$$

$$\mathbf{M} = \beta_1 \mathbf{A} \mathrm{tr}\, \boldsymbol{\kappa}_\parallel + \beta_2 \boldsymbol{\kappa}_\parallel^T + \beta_3 \boldsymbol{\kappa}_\parallel + \beta_4 \boldsymbol{\kappa}\cdot\mathbf{N}\otimes\mathbf{N}. \tag{D.27}$$

Supplemented with Eqs. (D.23) and (D.24), the linear constitutive equations (D.26), (D.27) constitute the setup of the linear boundary value problem with respect to the fields of translations and rotations. It describes micropolar shell equilibrium when the strains are infinitesimal.

Existence and uniqueness of weak solutions to the boundary value problems of statics and dynamics of micropolar linear shells as well as some spectral properties of the shells were established in [54].

D.6 Constitutive Restrictions for Micropolar Shells

As in 3D elasticity, in the theory of micropolar shells we should supplement the equilibrium/motion equations with constitutive equations. We will do that in the frame of general nonlinear shell theory similarly to what was done in 3D elasticity. Following [55] we will formulate the generalized Coleman-Noll inequality (GCN-condition), the strong ellipticity condition for the equilibrium equations and the Hadamard inequality. The inequalities impose some restrictions on constitutive equations of elastic shells under finite deformation. We also will prove that the Coleman-Noll inequality implies strong ellipticity of shell equilibrium equations. We begin with the linear theory.

D.6.1 Linear Theory of Micropolar Shell

Suppose that specific strain energy $W(\boldsymbol{\varepsilon}, \boldsymbol{\kappa})$ is positive definite. W is a quadratic form depending on the linear strain tensor and linear bending strain tensor. For an isotropic shell, W takes the form (D.25). Positivity of the quadratic form (D.25) with respect to $\boldsymbol{\varepsilon}$ and $\boldsymbol{\kappa}$ is equivalent to the following set of inequalities

$$2\alpha_1 + \alpha_2 + \alpha_3 > 0, \quad \alpha_2 + \alpha_3 > 0, \quad \alpha_3 - \alpha_2 > 0, \quad \alpha_4 > 0, \tag{D.28}$$
$$2\beta_1 + \beta_2 + \beta_3 > 0, \quad \beta_2 + \beta_3 > 0, \quad \beta_3 - \beta_2 > 0, \quad \beta_4 > 0.$$

The inequality

$$W(\boldsymbol{\varepsilon}, \boldsymbol{\kappa}) > 0, \quad \forall \boldsymbol{\varepsilon}, \boldsymbol{\kappa} \neq \mathbf{0}$$

and the inequalities for elastic constants of isotropic material (D.28) that are its consequences, present the simplest example of additional inequalities in the shell theory. If the inequalities fail this leads to a number of pathological consequences. For example boundary value problems of linear shell theory can have few solutions or can have no solution for some loads. Next, the propagation of waves in some directions becomes impossible that is not natural from the physical point of view. Note that for finite strains, the positive definiteness of specific energy $W(\mathbf{E}, \mathbf{K})$ is not a warranty that the desired properties of boundary value problems or wave propagation hold, here we should introduce some additional restrictions.

D.6.2 Coleman–Noll Inequality for Elastic Shells

Suppose a solution of equilibrium problem for a nonlinear elastic shell of Cosserat type is known. Let us call it the initial or basic stressed state. The state is defined by vector field $\boldsymbol{\rho}(q^\alpha)$ and tensor field $\mathbf{Q}(q^\alpha)$. Now we consider some equilibrium shell state that perturbs the basic state. If the difference between the state is infinitesimal we can linearize the equations with respect to the quantities characterizing the difference between the states. Let us denote the small increment of various quantities characterizing the perturbed equilibrium with the dot superscript like $\mathbf{D}^{\cdot}$. This quantity can be calculated by the formula:

$$\mathbf{D}^{\cdot} = \frac{\mathrm{d}}{\mathrm{d}\tau}\mathbf{D}[\nabla_s(\boldsymbol{\rho} + \tau\mathbf{u}, \mathbf{Q} - \tau\mathbf{Q}\times\boldsymbol{\theta}, \nabla_s(\mathbf{Q} - \tau\mathbf{Q}\times\boldsymbol{\theta}))]\Big|_{\tau=0}, \tag{D.29}$$

where $\mathbf{u}$ is the vector of additional infinitesimal translation and $\boldsymbol{\theta}$ is the vector of additional infinitesimal rotation characterizing the small rotation with respect to the initial stressed state. The following relations are valid

$$\boldsymbol{\rho}^{\cdot} = \mathbf{u}, \quad \mathbf{Q}^{\cdot} = -\mathbf{Q}\times\boldsymbol{\theta}, \quad \mathbf{E}^{\cdot} = \mathbf{F}\cdot\boldsymbol{\varepsilon}\cdot\mathbf{Q}^T, \quad \mathbf{K}^{\cdot} = \mathbf{F}\cdot\boldsymbol{\kappa}\cdot\mathbf{Q}^T, \tag{D.30}$$

$$\boldsymbol{\varepsilon} = \nabla_s\mathbf{u} + \mathbf{A}\times\boldsymbol{\theta}, \quad \boldsymbol{\kappa} = \nabla_s\boldsymbol{\theta}, \tag{D.31}$$

where $\boldsymbol{\varepsilon}$ and $\boldsymbol{\kappa}$ are the linear stretch tensor and linear wryness tensor introduced in (D.21).

As a reference configuration it can be chosen any stressed shell state. To avoid awkward expressions and to simplify the calculations, we assume the reference configuration to be the initial (basic) stressed state of the shell. This means that in the reference configuration $\mathbf{F} = \mathbf{E} = \mathbf{I} - \mathbf{N}\otimes\mathbf{N}$, $\mathbf{Q} = \mathbf{I}$, $\mathbf{K} = \mathbf{O}$. Under this choice, using Eqs. (D.7), (D.10), (D.29)–(D.31) we have

$$\mathbf{D}^{\cdot} = \frac{\partial^2 W}{\partial\mathbf{E}\partial\mathbf{E}}\bullet\boldsymbol{\varepsilon} + \frac{\partial^2 W}{\partial\mathbf{E}\partial\mathbf{K}}\bullet\boldsymbol{\kappa} - \mathbf{T}\times\boldsymbol{\theta}, \quad \mathbf{G}^{\cdot} = \frac{\partial^2 W}{\partial\mathbf{K}\partial\mathbf{E}}\bullet\boldsymbol{\varepsilon} + \frac{\partial^2 W}{\partial\mathbf{K}\partial\mathbf{K}}\bullet\boldsymbol{\kappa} - \mathbf{M}\times\boldsymbol{\theta}. \tag{D.32}$$

Suppose that in the initial and perturbed stressed shell states the external couples are zero $\mathbf{m} = \boldsymbol{\mu} = \mathbf{0}$ and the external forces are "dead". Then the total potential energy of the shell is

$$\Pi = \iint\limits_{\Sigma} W\,\mathrm{d}\Sigma - \iint\limits_{\Sigma} \mathbf{f}\cdot(\boldsymbol{\rho} - \mathbf{P})\mathrm{d}\Sigma - \int\limits_{\omega_2} \mathbf{t}\cdot(\boldsymbol{\rho} - \mathbf{P})\mathrm{d}s.$$

Let us consider the energy increment for the perturbed equilibrium state with respect to the initial energy taking into account the members of the second order of smallness

$$\Pi - \Pi_0 = \tau\left(\frac{\mathrm{d}\Pi}{\mathrm{d}\tau}\right)_{\tau=0} + \frac{1}{2}\tau^2\left(\frac{\mathrm{d}^2\Pi}{\mathrm{d}\tau^2}\right)_{\tau=0} + \cdots$$

By the constitutive relations (D.7) and Eqs. (D.30), (D.31), we get

$$\frac{\mathrm{d}\Pi}{\mathrm{d}\tau} = \iint\limits_{\Sigma} \left[\mathrm{tr}\left(\mathbf{D}^T\cdot\nabla_s\mathbf{u}\right) + \mathrm{tr}\left(\mathbf{D}^T\cdot\mathbf{F}\times\boldsymbol{\theta}\right) + \mathrm{tr}\left(\mathbf{G}^T\cdot\nabla_s\boldsymbol{\theta}\right)\right]\mathrm{d}\Sigma \\ - \iint\limits_{\Sigma} \mathbf{f}\cdot\mathbf{u}\,\mathrm{d}\Sigma - \int\limits_{\omega_2} \boldsymbol{\varphi}\cdot\mathbf{u}\,\mathrm{d}s. \tag{D.33}$$

We recall that the basic stressed shell state is the reference configuration of the shell. Differentiating Eq. (D.33) with respect to parameter τ and using Eqs. (D.30) we obtain

$$\left.\frac{\mathrm{d}^2\Pi}{\mathrm{d}\tau^2}\right|_{\tau=0} = \iint\limits_{\Sigma} \left[\mathrm{tr}\left(\mathbf{D}^{\cdot T}\cdot\nabla_s\mathbf{u}\right) + \mathrm{tr}\left(\mathbf{D}^{\cdot T}\times\boldsymbol{\theta}\right)\right. \\ \left. + \mathrm{tr}\left(\mathbf{T}^T\cdot(\nabla_s\mathbf{u})\times\boldsymbol{\theta}\right) + \mathrm{tr}\left(\mathbf{G}^{\cdot T}\cdot\boldsymbol{\kappa}\right)\right]\mathrm{d}\Sigma.$$

As we have chosen equilibrium of shell as the basic state, with use of Eqs. (D.6) and (D.8) we get that the first variation of the energy vanishes

$$\left.\frac{\mathrm{d}\Pi}{\mathrm{d}\tau}\right|_{\tau=0} = 0.$$

By Eqs. (D.31) and (D.32), the second energy variation takes the form

$$\left.\frac{\mathrm{d}^2\Pi}{\mathrm{d}\tau^2}\right|_{\tau=0} = 2\iint\limits_{\Sigma} w\,\mathrm{d}\Sigma, \quad w = w' + w'', \tag{D.34}$$

where

$$w' = \frac{1}{2}\boldsymbol{\varepsilon}\bullet\frac{\partial^2 W}{\partial\mathbf{E}\partial\mathbf{E}}\bullet\boldsymbol{\varepsilon} + \boldsymbol{\varepsilon}\bullet\frac{\partial^2 W}{\partial\mathbf{E}\partial\mathbf{K}}\bullet\boldsymbol{\kappa} + \frac{1}{2}\boldsymbol{\kappa}\bullet\frac{\partial^2 W}{\partial\mathbf{K}\partial\mathbf{K}}\bullet\boldsymbol{\kappa}, \\ w'' = \mathrm{tr}\left(\boldsymbol{\theta}\times\mathbf{T}^T\cdot\boldsymbol{\varepsilon}\right) - \frac{1}{2}\mathrm{tr}\left(\boldsymbol{\theta}\times\mathbf{T}^T\times\boldsymbol{\theta}\right) + \frac{1}{2}\mathrm{tr}\left(\boldsymbol{\theta}\times\mathbf{M}^T\cdot\boldsymbol{\kappa}\right). \tag{D.35}$$

w is the increment of the elastic energy of the initially prestressed shell under additional infinitesimal deformations. By Eqs. (D.34) and (D.35), this incremental energy splits into two parts: the pure strain energy, w', and the energy of rotations w''. The coefficients in the quadratic form w'' are expressed in terms of the stress and couple stress tensors of the initially prestressed state, they do not depend on the properties of shell material. If the basic stressed state of the shell is natural, that is $\mathbf{T} = \mathbf{M} = \mathbf{0}$, then $w = w'$ and the energy is a quadratic form of tensors $\boldsymbol{\varepsilon}$ and $\boldsymbol{\kappa}$. It is easily seen that the decomposition (D.34) and Eqs. (D.35) coincide with the corresponding quantities for the increment of the strain energy density of 3D micropolar body [56] up to the notation.

The Coleman–Noll constitutive inequality is one of well-known in nonlinear elasticity [8, 57, 58]. Its differential form, a so-called GCN-condition, expresses the property that for any reference configuration, the increment of the elastic energy density for arbitrary infinitesimal non-zero strains should be positive. Note that the Coleman–Noll inequality in 3D elasticity does not restrict the constitutive equations with respect to the rotations.

Taking into account the decomposition (D.34) of the energy we obtain an analogue of the *Coleman–Noll inequality* for micropolar elastic shells

$$w'(\boldsymbol{\varepsilon}, \boldsymbol{\kappa}) > 0 \quad \forall \boldsymbol{\varepsilon} \neq \mathbf{0}, \quad \boldsymbol{\kappa} \neq \mathbf{0}. \tag{D.36}$$

Using Eqs. (D.35) we rewrite (D.36) in the equivalent form

$$\frac{\mathrm{d}^2}{\mathrm{d}\tau^2} W(\mathbf{E} + \tau\boldsymbol{\varepsilon}, \mathbf{K} + \tau\boldsymbol{\kappa})\Big|_{\tau=0} > 0 \quad \forall \boldsymbol{\varepsilon} \neq \mathbf{0}, \quad \boldsymbol{\kappa} \neq \mathbf{0}. \tag{D.37}$$

Condition (D.37) satisfies the principle of material frame-indifference, it can serve as a constitutive inequality for elastic shells.

D.6.3 Strong Ellipticity and Hadamard Inequality

In nonlinear elasticity, the strong ellipticity condition and its weak form, the Hadamard inequality, are other known constitutive restrictions. Following the partial differential equations theory (PDE) [59–61] we formulate the strong ellipticity condition of the equilibrium equations (D.6). For dead loads, the linearized equilibrium equations are

$$\nabla_s \cdot \mathbf{D}^{\cdot} = \mathbf{0}, \quad \nabla_s \cdot \mathbf{G}^{\cdot} + \left[\mathbf{F}^T \cdot \mathbf{D}^{\cdot} + (\nabla_s \mathbf{u})^T \cdot \mathbf{D}\right]_\times = \mathbf{0}, \tag{D.38}$$

where $\mathbf{D}^{\cdot}$ and $\mathbf{G}^{\cdot}$ are defined by the formulae similar to (D.29). Equations (D.38) constitute a system of linear PDE of second order with respect to $\mathbf{u}$ and $\boldsymbol{\theta}$. The second order parts of the differential operators in Eqs. (D.38) are

$$\nabla_s \cdot \left\{ \left[\frac{\partial^2 W}{\partial \mathbf{E} \partial \mathbf{E}} \bullet \left((\nabla_s \mathbf{u}) \cdot \mathbf{Q}^T \right) + \frac{\partial^2 W}{\partial \mathbf{E} \partial \mathbf{K}} \bullet \left((\nabla_s \boldsymbol{\theta}) \cdot \mathbf{Q}^T \right) \right] \cdot \mathbf{Q} \right\},$$

$$\nabla_s \cdot \left\{ \left[\frac{\partial^2 W}{\partial \mathbf{K} \partial \mathbf{E}} \bullet \left((\nabla_s \mathbf{u}) \cdot \mathbf{Q}^T \right) + \frac{\partial^2 W}{\partial \mathbf{K} \partial \mathbf{K}} \bullet \left((\nabla_s \boldsymbol{\theta}) \cdot \mathbf{Q}^T \right) \right] \cdot \mathbf{Q} \right\}.$$

Now we can formulate the condition of strong ellipticity for system (D.38). Following a formal procedure from [59], we replace the differential operators ∇_s by the unit vector $\boldsymbol{\nu}$ tangential to surface Σ and vector fields $\mathbf{u}$ and $\boldsymbol{\theta}$ by vectors $\mathbf{a}$ and $\mathbf{b}$, respectively. Thus, we get the algebraic expressions

$$\boldsymbol{\nu} \cdot \left\{ \left[\frac{\partial^2 W}{\partial \mathbf{E} \partial \mathbf{E}} \bullet \left(\boldsymbol{\nu} \otimes \mathbf{a} \cdot \mathbf{Q}^T \right) + \frac{\partial^2 W}{\partial \mathbf{E} \partial \mathbf{K}} \bullet \left(\boldsymbol{\nu} \otimes \mathbf{b} \cdot \mathbf{Q}^T \right) \right] \cdot \mathbf{Q} \right\},$$

$$\boldsymbol{\nu} \cdot \left\{ \left[\frac{\partial^2 W}{\partial \mathbf{K} \partial \mathbf{E}} \bullet \left(\boldsymbol{\nu} \otimes \mathbf{a} \cdot \mathbf{Q}^T \right) + \frac{\partial^2 W}{\partial \mathbf{K} \partial \mathbf{K}} \bullet \left(\boldsymbol{\nu} \otimes \mathbf{b} \cdot \mathbf{Q}^T \right) \right] \cdot \mathbf{Q} \right\}.$$

Multiply the first equation by vector **a**, the second equation by **b** and add the results. Then we get the strong ellipticity condition of Eqs. (D.38):

$$\boldsymbol{\nu}\cdot\left\{\left[\frac{\partial^2 W}{\partial\mathbf{E}\partial\mathbf{E}}\bullet\left(\boldsymbol{\nu}\otimes\mathbf{a}\cdot\mathbf{Q}^T\right)+\frac{\partial^2 W}{\partial\mathbf{E}\partial\mathbf{K}}\bullet\left(\boldsymbol{\nu}\otimes\mathbf{b}\cdot\mathbf{Q}^T\right)\right]\cdot\mathbf{Q}\right\}\cdot\mathbf{a}$$
$$+\boldsymbol{\nu}\cdot\left\{\left[\frac{\partial^2 W}{\partial\mathbf{K}\partial\mathbf{E}}\bullet\left(\boldsymbol{\nu}\otimes\mathbf{a}\cdot\mathbf{Q}^T\right)+\frac{\partial^2 W}{\partial\mathbf{K}\partial\mathbf{K}}\bullet\left(\boldsymbol{\nu}\otimes\mathbf{b}\cdot\mathbf{Q}^T\right)\right]\cdot\mathbf{Q}\right\}\cdot\mathbf{b}>0,\quad\forall\mathbf{a},\mathbf{b}\neq\mathbf{0}.$$

Using the operation $\bullet$, we transform the inequality as follows

$$\left(\boldsymbol{\nu}\otimes\mathbf{a}\cdot\mathbf{Q}^T\right)\bullet\frac{\partial^2 W}{\partial\mathbf{E}\partial\mathbf{E}}\bullet\left(\boldsymbol{\nu}\otimes\mathbf{a}\cdot\mathbf{Q}^T\right)+2\left(\boldsymbol{\nu}\otimes\mathbf{a}\cdot\mathbf{Q}^T\right)\bullet\frac{\partial^2 W}{\partial\mathbf{E}\partial\mathbf{K}}\bullet\left(\boldsymbol{\nu}\otimes\mathbf{b}\cdot\mathbf{Q}^T\right)$$
$$+\left(\boldsymbol{\nu}\otimes\mathbf{b}\cdot\mathbf{Q}^T\right)\bullet\frac{\partial^2 W}{\partial\mathbf{K}\partial\mathbf{E}}\bullet\left(\boldsymbol{\nu}\otimes\mathbf{b}\cdot\mathbf{Q}^T\right)>0,\qquad\forall\mathbf{a},\mathbf{b}\neq\mathbf{0}.$$

In matrix notations, we rewrite this in a compact form

$$\boldsymbol{\xi}\cdot\mathbb{A}(\boldsymbol{\nu})\cdot\boldsymbol{\xi}>0,\quad\forall\boldsymbol{\nu}\neq\mathbf{0},\quad\boldsymbol{\nu}\cdot\mathbf{N}=0,\quad\forall\boldsymbol{\xi}\in\mathbb{R}^6,\quad\boldsymbol{\xi}\neq\mathbf{0},\qquad\text{(D.39)}$$

where $\boldsymbol{\xi}=(\mathbf{a}',\mathbf{b}')\in\mathbb{R}^6$, $\mathbf{a}'=\mathbf{a}\cdot\mathbf{Q}^T$, $\mathbf{b}'=\mathbf{b}\cdot\mathbf{Q}^T$, and matrix $\mathbb{A}(\boldsymbol{\nu})$ is

$$\mathbb{A}(\boldsymbol{\nu})\triangleq\begin{bmatrix}\dfrac{\partial^2 W}{\partial\mathbf{E}\partial\mathbf{E}}\{\boldsymbol{\nu}\} & \dfrac{\partial^2 W}{\partial\mathbf{E}\partial\mathbf{K}}\{\boldsymbol{\nu}\}\\ \dfrac{\partial^2 W}{\partial\mathbf{K}\partial\mathbf{E}}\{\boldsymbol{\nu}\} & \dfrac{\partial^2 W}{\partial\mathbf{K}\partial\mathbf{K}}\{\boldsymbol{\nu}\}\end{bmatrix},$$

where for any fourth-order tensor $\mathbf{K}$ and vector $\boldsymbol{\nu}$ we denote

$$\mathbf{K}\{\boldsymbol{\nu}\}\triangleq K_{klmn}\nu_k\nu_m\mathbf{i}_l\otimes\mathbf{i}_n.$$

Inequality (D.39) is the *strong ellipticity condition*of the equilibrium equations (D.6) for the elastic shell. A weak form of inequality (D.39) is an analogue of the *Hadamard inequality*. These inequalities are examples of possible restrictions of the constitutive equations of elastic shells under finite deformations. As for the theory of simple materials, a failure in inequality (D.39) can lead to the existence of non-smooth solutions to equilibrium equations (D.6).

The strong ellipticity condition can be written in the equivalent form

$$\left.\frac{\mathrm{d}^2}{\mathrm{d}\tau^2}W(\mathbf{E}+\tau\boldsymbol{\nu}\otimes\mathbf{a}',\mathbf{K}+\tau\boldsymbol{\nu}\otimes\mathbf{b}')\right|_{\tau=0}>0\quad\forall\boldsymbol{\nu},\mathbf{a}',\mathbf{b}'\neq\mathbf{0}.\qquad\text{(D.40)}$$

Comparing the strong ellipticity condition (D.40) and the Coleman–Noll inequality (D.37) one can see that the latter implies the former. Indeed, inequality (D.37) holds for any tensors $\boldsymbol{\varepsilon}$ and $\boldsymbol{\kappa}$. Note that $\boldsymbol{\varepsilon}$ and $\boldsymbol{\kappa}$ may be nonsymmetric tensors, in general. Substituting relations $\boldsymbol{\varepsilon}=\boldsymbol{\nu}\otimes\mathbf{a}'$ and $\boldsymbol{\kappa}=\boldsymbol{\nu}\otimes\mathbf{b}'$ to inequality (D.37), we immediately obtain inequality (D.40). Thus, the strong ellipticity condition is a particular case of the Coleman–Noll inequality. We watch an essential difference between the micropolar shell theory and the theory of simple elastic materials [8, 57]: in the latter these two properties are independent in the sense that neither of them implies the other.

In the shell theory, the following particular constitutive relation is widely used

$$W(\mathbf{E},\mathbf{K}) = W_1(\mathbf{E}) + W_2(\mathbf{K}). \tag{D.41}$$

For example, Eq. (D.19) has the form of (D.41). Now condition (D.39) is equivalent to two simpler inequalities

$$\mathbf{a}\cdot\frac{\partial^2 W_1}{\partial\mathbf{E}\partial\mathbf{E}}\{\boldsymbol{\nu}\}\cdot\mathbf{a} > 0, \quad \mathbf{b}\cdot\frac{\partial^2 W_2}{\partial\mathbf{K}\partial\mathbf{K}}\{\boldsymbol{\nu}\}\cdot\mathbf{b} > 0.$$

As an example, let us consider consequences of conditions (D.39) for constitutive equation (D.19). In this case we have

$$\begin{aligned}\frac{\partial^2 W_1}{\partial\mathbf{E}\partial\mathbf{E}}\{\boldsymbol{\nu}\} &= \alpha_3\mathbf{A} + (\alpha_1+\alpha_2)\boldsymbol{\nu}\otimes\boldsymbol{\nu} + \alpha_4\mathbf{N}\otimes\mathbf{N},\\ \frac{\partial^2 W_2}{\partial\mathbf{K}\partial\mathbf{K}}\{\boldsymbol{\nu}\} &= \beta_3\mathbf{A} + (\beta_1+\beta_2)\boldsymbol{\nu}\otimes\boldsymbol{\nu} + \beta_4\mathbf{N}\otimes\mathbf{N}.\end{aligned} \tag{D.42}$$

Now inequality (D.39) is valid under the following conditions

$$\alpha_3 > 0, \quad \alpha_1+\alpha_2+\alpha_3 > 0, \quad \alpha_4 > 0, \quad \beta_3 > 0, \quad \beta_1+\beta_2+\beta_3 > 0, \quad \beta_4 > 0. \tag{D.43}$$

For a linear isotropic shell, inequalities (D.43) provide the strong ellipticity of equilibrium equations (D.23), they are weaker than the conditions of positive definiteness (D.28). Considering the constitutive equations of an isotropic micropolar shell (D.39) we have reduced inequality (D.39) to the inequalities (D.43).

D.6.4 Strong Ellipticity Condition and Acceleration Waves

Using the approach of [55, 62, 63], we will show that inequality (D.39) coincides with the conditions of the propagation of acceleration waves in a shell. We consider a motion when on a smooth curve $\mathscr{C}(t) \subset \Sigma$ called *singular*, continuous kinematic and dynamic quantities can jump, see Fig. D.1. We assume that the limit values of these quantities exist on $\mathscr{C}$ being different from the opposite sides of $\mathscr{C}$ in general. The jump of quantity ψ on $\mathscr{C}$ will be denoted by the double brackets: $[\![\psi]\!] = \psi^+ - \psi^-$, where $\psi^\pm$ are one-side limits of ψ.

An *acceleration wave* (a *weak-discontinuity wave* or *second-order singular curve*) is a moving singular curve $\mathscr{C}$ on which the second derivatives of the radius-vector $\boldsymbol{\rho}$ and the microrotation tensor $\mathbf{Q}$ with respect to the spatial coordinates and time are discontinuous, while $\boldsymbol{\rho}$, $\mathbf{Q}$ and their first derivatives are continuous that means that on $\mathscr{C}$

$$[\![\mathbf{F}]\!] = \mathbf{0}, \quad [\![\nabla_s \mathbf{Q}]\!] = \mathbf{0}, \quad [\![\mathbf{v}]\!] = \mathbf{0}, \quad [\![\boldsymbol{\omega}]\!] = \mathbf{0}. \tag{D.44}$$

By Eqs. (D.2), the stretch measure $\mathbf{E}$ and the wryness tensor $\mathbf{K}$ are continuous on $\mathscr{C}$. By constitutive equations (D.7), the jumps of tensors $\mathbf{D}$ and $\mathbf{G}$ are absent. Applying the Maxwell theorem [57, 58] to continuous fields of velocities $\mathbf{v}$ and $\boldsymbol{\omega}$, surface stress tensor $\mathbf{D}$, and the surface couple stress tensor $\mathbf{G}$, we deduce a system of equations that relates the jumps of the derivatives of these quantities with respect to the spatial coordinates and time

$$\left[\!\!\left[\frac{d\mathbf{v}}{dt}\right]\!\!\right] = -V\mathbf{a}, \quad [\![\nabla_s \mathbf{v}]\!] = \boldsymbol{\nu} \otimes \mathbf{a}, \quad \left[\!\!\left[\frac{d\boldsymbol{\omega}}{dt}\right]\!\!\right] = -V\mathbf{b}, \quad [\![\nabla_s \boldsymbol{\omega}]\!] = \boldsymbol{\nu} \otimes \mathbf{b}, \tag{D.45}$$

$$V[\![\nabla_s \cdot \mathbf{D}]\!] = -\boldsymbol{\nu} \cdot \left[\!\!\left[\frac{d\mathbf{D}}{dt}\right]\!\!\right], \quad V[\![\nabla_s \cdot \mathbf{G}]\!] = -\boldsymbol{\nu} \cdot \left[\!\!\left[\frac{d\mathbf{G}}{dt}\right]\!\!\right].$$

Here $\mathbf{a}$ and $\mathbf{b}$ are the vectorial amplitudes of the jumps of the linear and angular accelerations, respectively, $\boldsymbol{\nu}$ is the unit normal vector to $\mathscr{C}$ such that $\mathbf{n} \cdot \boldsymbol{\nu} = 0$, and V is the velocity of the surface $\mathscr{C}$ in the direction $\boldsymbol{\nu}$. If the external forces and couples are continuous, the relations

$$[\![\nabla_s \cdot \mathbf{D}]\!] = \rho \left[\!\!\left[\frac{d\mathbf{K}_1}{dt}\right]\!\!\right], \quad [\![\nabla_s \cdot \mathbf{G}]\!] = \rho\gamma \left[\!\!\left[\frac{d\mathbf{K}_2}{dt}\right]\!\!\right]$$

follow immediately from the motion equations (D.13).

Differentiating constitutive equations (D.7) and using Eqs. (D.44) and (D.45), we express the last relations in terms of vector amplitudes $\mathbf{a}$ and $\mathbf{b}$

$$\begin{aligned}
&\boldsymbol{\nu} \cdot \frac{\partial^2 W}{\partial \mathbf{E} \partial \mathbf{E}} \bullet \left(\boldsymbol{\nu} \otimes \mathbf{a} \cdot \mathbf{Q}^T\right) + \boldsymbol{\nu} \cdot \frac{\partial^2 W}{\partial \mathbf{E} \partial \mathbf{K}} \bullet \left(\boldsymbol{\nu} \otimes \mathbf{b} \cdot \mathbf{Q}^T\right) \\
&\qquad = \rho V^2 \left[\mathbf{a} \cdot \mathbf{Q}^T + \left(\mathbf{Q} \cdot \boldsymbol{\Theta}_1^T \cdot \mathbf{Q}^T\right) \cdot \left(\mathbf{b} \cdot \mathbf{Q}^T\right)\right], \\
&\boldsymbol{\nu} \cdot \frac{\partial^2 W}{\partial \mathbf{K} \partial \mathbf{E}} \bullet \left(\boldsymbol{\nu} \otimes \mathbf{a} \cdot \mathbf{Q}^T\right) + \boldsymbol{\nu} \cdot \frac{\partial^2 W}{\partial \mathbf{K} \partial \mathbf{K}} \bullet \left(\boldsymbol{\nu} \otimes \mathbf{b} \cdot \mathbf{Q}^T\right) \\
&\qquad = \rho V^2 \left[\left(\mathbf{Q} \cdot \boldsymbol{\Theta}_1 \cdot \mathbf{Q}^T\right) \cdot \left(\mathbf{a} \cdot \mathbf{Q}^T\right) + \left(\mathbf{Q} \cdot \boldsymbol{\Theta}_2 \cdot \mathbf{Q}^T\right) \cdot \left(\mathbf{b} \cdot \mathbf{Q}^T\right)\right].
\end{aligned}$$

Hence the strong ellipticity condition can be written in a compact form

$$\mathbb{A}(\boldsymbol{\nu}) \cdot \boldsymbol{\xi} = \rho V^2 \mathbb{B} \cdot \boldsymbol{\xi}, \qquad \mathbb{B} = \begin{bmatrix} \mathbf{I} & \mathbf{Q} \cdot \boldsymbol{\Theta}_1^T \cdot \mathbf{Q}^T \\ \mathbf{Q} \cdot \boldsymbol{\Theta}_1 \cdot \mathbf{Q}^T & \mathbf{Q} \cdot \boldsymbol{\Theta}_2 \cdot \mathbf{Q}^T \end{bmatrix}. \tag{D.46}$$

Thus, the problem of propagation of an acceleration wave in a shell is reduced to the spectral problem given by algebraic equations (D.46). Existence of potential-energy function W implies that $\mathbb{A}(\boldsymbol{\nu})$ is symmetric. Matrix $\mathbb{B}$ is also symmetric and positive definite. This enables us to formulate an analogue of the *Fresnel–Hadamard–Duhem theorem* for the elastic shell:

Theorem D.1 *In an elastic shell, for any propagation direction specified by vector* $\mathbf{v}$, *the squared velocities of a second order singular curve (the acceleration wave) are real.*

Note that positive definiteness of $\mathbb{A}(\mathbf{v})$, which is necessary and sufficient for the wave velocity V to be real, coincides with the strong ellipticity inequality (D.39).

For a physically linear shell, we present an example of solution of the problem (D.46). Let $\boldsymbol{\Theta}_1$ be zero and $\boldsymbol{\Theta}_2$ be a spherical part of tensor (ball tensor), that is $\boldsymbol{\Theta}_2 = j\mathbf{I}$, where j is the rotatory inertia measure. Let the inequalities (D.43) hold. Then the solutions of Eq. (D.46) are

$$V_1 = \sqrt{\frac{\alpha_3}{\rho}}, \quad \xi_1 = (\boldsymbol{\tau}, \mathbf{0}), \quad V_2 = \sqrt{\frac{\alpha_1 + \alpha_2 + \alpha_3}{\rho}}, \quad \xi_2 = (\mathbf{v}, \mathbf{0}), \tag{D.47}$$

$$V_3 = \sqrt{\frac{\alpha_4}{\rho}}, \quad \xi_3 = (\mathbf{N}, \mathbf{0}), \quad V_4 = \sqrt{\frac{\beta_3}{\rho j}}, \quad \xi_4 = (\mathbf{0}, \boldsymbol{\tau}),$$

$$V_5 = \sqrt{\frac{\beta_1 + \beta_2 + \beta_3}{\rho j}}, \quad \xi_5 = (\mathbf{0}, \mathbf{v}), \quad V_6 = \sqrt{\frac{\beta_4}{\rho j}}, \quad \xi_6 = (\mathbf{0}, \mathbf{N}).$$

The solutions (D.47) are similar to the 3D case (5.30) and (5.31) describe the *transverse and longitudinal acceleration waves* and *transverse and longitudinal acceleration waves of microrotation*, respectively.

D.6.5 Ordinary Ellipticity

If the equilibrium equations are not elliptic the continuity of solutions can fail. Let us consider this in more detail. We will assume the singular curves to be time-independent. Suppose on the shell surface Σ there exists a curve $\mathscr{C}$ on which there happen a jump in the values of second derivatives of position vector $\boldsymbol{\rho}$ or microrotation tensor $\mathbf{Q}$. Such a jump will be called the *weak discontinuity*. As the curvature of Σ is determined through second derivatives of $\boldsymbol{\rho}$, such discontinuity can be exhibited as wrinkling of the shell surface.

From the equilibrium equations it follows $[\![\nabla_s \cdot \mathbf{D}]\!] = \mathbf{0}$, $[\![\nabla_s \cdot \mathbf{G}]\!] = \mathbf{0}$. Repeating the transformations of the previous section, we transform these to

$$\mathbb{A}(\mathbf{v}) \cdot \xi = \mathbf{0}, \quad \xi = (\mathbf{a}', \mathbf{b}') \in \mathbb{R}^6. \tag{D.48}$$

Existence of nontrivial solutions of Eq. (D.48) means that the weak discontinuities arise. The nontrivial solutions exist if the determinant of matrix $\mathbb{A}(\mathbf{v})$ is zero. If

$$\det \mathbb{A}(\mathbf{v}) \neq 0, \tag{D.49}$$

the weak discontinuities are impossible.

For the constitutive relation $W = W_1(\mathbf{E}) + W_2(\mathbf{K})$, condition (D.49) splits into two conditions

$$\det \frac{\partial^2 W_1}{\partial \mathbf{E} \partial \mathbf{E}} \{v\} \neq 0, \quad \det \frac{\partial^2 W_2}{\partial \mathbf{K} \partial \mathbf{K}} \{v\} \neq 0. \tag{D.50}$$

As an example, we consider conditions (D.50) for the constitutive relations of a physically linear shell (D.19). Using Eqs. (D.42) we can show that conditions (D.50) reduce to the inequalities

$$\alpha_3 \neq 0, \quad \alpha_1 + \alpha_2 + \alpha_3 \neq 0, \quad \alpha_4 \neq 0, \quad \beta_3 \neq 0, \quad \beta_1 + \beta_2 + \beta_3 \neq 0, \quad \beta_4 \neq 0.$$

Condition (D.49) is the *ellipticity condition* of the equilibrium equations of shell theory (ellipticity in the Petrovsky sense). The condition follows from the general definition of ellipticity in PDE theory [60, 64, 65]. Condition (D.49) is also called the *ordinary ellipticity condition*, it is weaker than the strong ellipticity condition (D.39).

References

1. L.P. Lebedev, M.J. Cloud, V.A. Eremeyev, *Tensor Analysis with Applications in Mechanics* (World Scientific, Hackensack, 2010)
2. W. Pietraszkiewicz, V.A. Eremeyev, On vectorially parameterized natural strain measures of the non-linear Cosserat continuum. Int. J. Solids Struct. **46**(11–12), 2477–2480 (2009)
3. R.P. Feynman, R.B. Leighton, M. Sands, *The Feynman Lectures on Physics*, vol. 1, 6th edn. (Addison-Wesley, Reading, 1977)
4. G.A. Korn, T.M. Korn, *Mathematical Handbook for Scientists and Engineers: Definitions, Theorems and Formulas for Reference and Review*, 2nd edn. (McGraw-Hill, New York, 1968)
5. J.F. Nye, *Physical Properties of Crystals. Their Representation by Tensors and Matrices* (Clarendon Press, Oxford, 1957)
6. J.P. Boehler (eds), *Applications of Tensor Functions in Solid Mechanics*. CISM Courses and Lectures, vol. 292. (Springer, Wien, 1987)
7. A.J.M. Spencer, Theory of invariants. In: A.C. Eringen (eds) *Continuum Physics*, vol. 1 (Academic Press, New-York, 1971), pp. 239–353
8. C. Truesdell, W. Noll, The nonlinear field theories of mechanics. In: S. Flügge (eds) *Handbuch der Physik*, vol. III/3. (Springer, Berlin, 1965), pp. 1–602
9. Q.S. Zheng, Theory of representations for tensor functions—a unified invariant approach to constitutive equations. Appl. Mech. Rev. **47**(11), 545–587 (1994)
10. M.M. Smith, R.F. Smith, Irreducable expressions for isotropic functions of two tensors. Int. J. Eng. Sci. **19**(6), 811–817 (1971)
11. A.I. Lurie, *Analytical Mechanics* (Springer, Berlin, 2001)
12. V.I. Arnold, *Mathematical Methods of Classical Mechanics*, 2nd edn. (Springer, New York, 1989)
13. C. Toupin, R. Truesdell, The classical field theories. In: S. Flügge (eds) *Handbuch der Physik*, vol. III/1. (Springer, Berlin, 1960), pp. 226–793

14. Hodges, D.H., *Nonlinear Composite Beam Theory, Progress in Astronautics and Aeronautics*, vol. 213 (American Institute of Aeronautics and Astronautics, Inc., Reston, 2006)
15. L. Librescu, O. Song, *Thin-Walled Composite Beams: Theory and Applications, Solid Mechanics and Its Applications*, vol. 131 (Springer, Dordrecht, 2006)
16. E. Cosserat, F. Cosserat, *Théorie des corps déformables* (Herman et Fils, Paris, 1909)
17. J.L. Ericksen, C. Truesdell, Exact tbeory of stress and strain in rods and shells. Arch. Ration. Mech. Anal. **1**(1), 295–323 (1958)
18. A.E. Green, P.M. Naghdi, Non-isothermal theory of rods, plates and shells. Int. J. Solids Struct. **6**, 209–244 (1970)
19. C.B. Kafadar, On the nonlinear theory of rods. Int. J. Eng. Sci. **10**(4), 369–391 (1972)
20. A.E. Green, P.M. Naghdi, M.L. Wenner, On the theory of rods. II. Developments by direct approach. Int. J. Solids Struct. **337**(1611), 485–507 (1974)
21. S.S. Antman, *Nonlinear Problems of Elasticity*, 2nd edn. (Springer Science Media, New York, 2005)
22. M.B. Rubin, *Cosserat Theories: Shells, Rods and Points* (Kluwer, Dordrecht, 2000)
23. V.V. Eliseev, *Mechanics of Elastic Bodies* (in Russian). (State Polytechnical University Publishing House, Petersburg, 1996)
24. V.A. Svetlitsky, *Statics of Rods* (Springer, Berlin, 2000)
25. J.G. Simmonds, A simple nonlinear thermodynamic theory of arbitrary elastic beams. J. Elast. **81**(1), 51–62 (2005)
26. D. Ieşan, *Classical and Generalized Models of Elastic Rods* (CRC Press, Boca Raton, 2009)
27. V.L. Berdichevsky, *Variational Principles of Continuum Mechanics. II. Applications* (Springer, Heidelberg, 2009)
28. Altenbach, H., Bîrsan, M., Eremeyev, V.A., On a thermodynamic theory of rods with two temperature fields. Acta Mechanica, doi: 10.1007/s00707-012-0632-1 (2012)
29. J. Altenbach, H. Altenbach, V.A. Eremeyev, On generalized Cosserat-type theories of plates and shells. A short review and bibliography. Arch. Appl. Mech. **80**(1), 73–92 (2010)
30. P.A. Zhilin, Mechanics of deformable directed surfaces. Int. J. Solids Struct. **12**(9–10), 635–648 (1976)
31. V.A. Eremeyev, Nonlinear micropolar shells: theory and applications. In: W. Pietraszkiewicz, C. Szymczak (eds) *Shell Structures: Theory and Applications* (Taylor and Francis, London, 2005), pp. 11–18
32. V.A. Eremeyev, L.M. Zubov, *Mechanics of Elastic Shells (in Russian)* (Nauka, Moscow, 2008)
33. L.M. Zubov, *Nonlinear Theory of Dislocations and Disclinations in Elastic Bodies* (Springer, Berlin, 1997)
34. E. Reissner, Linear and nonlinear theory of shells. In: Y.C. Fung, E.E. Sechler (eds) *Thin Shell Structures* (Prentice-Hall, Englewood Cliffs, 1974), pp. 29–44
35. A. Libai, J.G. Simmonds, Nonlinear elastic shell theory. Adv. Appl. Mech. **23**, 271–371 (1983)
36. A. Libai, J.G. Simmonds, *The Nonlinear Theory of Elastic Shells*, 2nd edn. (Cambridge University Press, Cambridge, 1998)
37. W. Pietraszkiewicz, Teorie nieliniowe powlok. In: C. Wozniak (ed.) *Mechanika sprezystych plyt ipowlok* (PWN, Warszawa, 2001), pp. 424–497
38. J. Chróścielewski, J. Makowski, W. Pietraszkiewicz, *Statics and Dynamics of Multyfolded Shells. Nonlinear Theory and Finite Elelement Method (in Polish)* (Wydawnictwo IPPT PAN, Warszawa, 2004)
39. J. Chróścielewski, W. Pietraszkiewicz, W. Witkowski, On shear correction factors in the non-linear theory of elastic shells. Int. J. Solids Struct. **47**(25–26), 3537–3545 (2010)
40. J. Chróścielewski, W. Witkowski, On some constitutive equations for micropolar plates. ZAMM **90**(1), 53–64 (2010)

41. J. Chróścielewski, W. Witkowski, FEM analysis of Cosserat plates and shells based on some constitutive relations. ZAMM **91**(5), 400–412 (2011)
42. V.A. Eremeyev, W. Pietraszkiewicz, The non-linear theory of elastic shells with phase transitions. J. Elast. **74**(1), 67–86 (2004)
43. V.A. Eremeyev, W. Pietraszkiewicz, Phase transitions in thermoelastic and thermoviscoelastic shells. Arch. Mech. **61**(1), 41–67 (2009)
44. V.A. Eremeyev, W. Pietraszkiewicz, Thermomechanics of shells undergoing phase transition. J. Mech. Phys. Solids **59**(7), 1395–1412 (2011)
45. W. Pietraszkiewicz, V.A. Eremeyev, V. Konopińska, Extended non-linear relations of elastic shells undergoing phase transitions. ZAMM **87**(2), 150–159 (2007)
46. J.G. Simmonds, The thermodynamical theory of shells: descent from 3-dimensions without thickness expansions. In: E.L. Axelrad, F.A. Emmerling (eds) *Flexible Shells, Theory and Applications* (Springer, Berlin, 1984), pp. 1–11
47. J.G. Simmonds, A classical, nonlinear thermodynamic theory of elastic shells based on a single constitutive assumption. J. Elast. **105**(1–2), 305–312 (2011)
48. W. Pietraszkiewicz, Refined resultant thermomechanics of shells. Int. J. Eng. Sci. **49**(10), 1112–1124 (2011)
49. V.A. Eremeyev, W. Pietraszkiewicz, Local symmetry group in the general theory of elastic shells. J. Elast. **85**(2), 125–152 (2006)
50. E. Reissner, On the theory of bending of elastic plates. J. Math. Phys. **23**, 184–194 (1944)
51. R.D. Mindlin, Influence of rotatory inertia and shear on flexural motions of isotropic elastic plates. Trans. ASME J. Appl. Mech. **18**, 31–38 (1951)
52. W. Pietraszkiewicz, Consistent second approximation to the elastic strain energy of a shell. ZAMM **59**, 206–208 (1979)
53. W. Pietraszkiewicz, *Finite Rotations and Langrangian Description in the Non-linear Theory of Shells* (Polish Scientific Publisher, Warsaw, 1979)
54. V.A. Eremeyev, L.P. Lebedev, Existence theorems in the linear theory of micropolar shells. ZAMM **91**(6), 468–476 (2011)
55. V.A. Eremeyev, L.M. Zubov, On constitutive inequalities in nonlinear theory of elastic shells. ZAMM **87**(2), 94–101 (2007)
56. V.A. Eremeyev, L.M. Zubov, On the stability of elastic bodies with couple stresses. Mech. Solids **29**(3), 172–181 (1994)
57. C. Truesdell, *A First Course in Rational Continuum Mechanics* (Academic Press, New York, 1977)
58. C. Truesdell, *Rational Thermodynamics*, 2nd edn. (Springer, New York, 1984)
59. J.L. Lions, E. Magenes, *Problèmes aux limites non homogènes et applications* (Dunod, Paris, 1968)
60. G. Fichera, Existence theorems in elasticity. In: S. Flügge (eds) *Handbuch der Physik*, vol. VIa/2 (Springer, Berlin, 1972), pp. 347–389
61. L. Hörmander, *Linear Partial Differential Equations*. A Series of Comprehensive Studies in Mathematics, vol. 116, 4th edn. (Springer, Berlin, 1976)
62. V.A. Eremeyev, Acceleration waves in micropolar elastic media. Doklady Phys. **50**(4), 204–206 (2005)
63. H. Altenbach, V.A. Eremeyev, L.P. Lebedev, L.A. Rendón, Acceleration waves and ellipticity in thermoelastic micropolar media. Arch. Appl. Mech. **80**(3), 217–227 (2010)
64. M. Agranovich, Elliptic boundary problems. In: M. Agranovich, Y. Egorov, M. Shubin (eds) *Partial Differential Equations IX: Elliptic Boundary Problems. Encyclopaedia of Mathematical Sciences*, vol. 79 (Springer, Berlin, 1997), pp 1–144
65. L. Nirenberg, *Topics in Nonlinear Functional Analysis* (American Mathematical Society, New York, 2001)

Index

V. A. Eremeyev et al., *Foundations of Micropolar Mechanics*,
SpringerBriefs in Continuum Mechanics,
DOI: 10.1007/978-3-642-28353-6, © The Author(s) 2013